Woodland offers birds an extremely wide range of habitats. In this book, the variation in woodland bird life and the factors that influence bird numbers and distributions are examined. What birds are found in which habitats? What effect does forestry and woodland management have? How can we enhance bird populations by habitat management? Are bird numbers and distributions in natural and managed forests different – and if so, why? The gamut of British woodland is covered from ancient coppice and wood-pasture in the lowlands, to recently-planted conifer forests in the uplands, and comparisons are drawn with mainland Europe and North America. The book discusses the effects of factors such as increased deer numbers, air pollution and the creation of new woods on lowland farms which are changing the face of our woodland today. This book will appeal to all those with an interest in woodland and its wildlife.

Bird life of woodland and forest

Bird Life Series

This series of books presents the bird life of the various habitats in the British Isles in a way that outlines how the birds are adapted to the various environments they inhabit and how the environments themselves shape the birds' behaviour and breeding patterns. They thus present a new, ecologically orientated approach to the understanding of the bird life of Britain and Ireland.

Other titles already published

Bird life of mountain and upland — Derek Ratcliffe
ISBN 0 521 33123 4
Bird life of coasts and estuaries — Peter Ferns
ISBN 0 521 34569 3

Forthcoming titles

Bird life of farmland, grassland and heathland
Bird life of freshwater wetlands
Bird life of towns, parks and gardens

The editor of the series is Professor C. M. Perrins, the Director of the Edward Grey Institute of Field Ornithology, University of Oxford.

Bird life of
woodland and forest

ROBERT J. FULLER

With line illustrations by Chris Rose

CAMBRIDGE
UNIVERSITY PRESS

PUBLISHED BY THE PRESS SYNDICATE OF THE UNIVERSITY OF CAMBRIDGE
The Pitt Building, Trumpington Street, Cambridge, United Kingdom

CAMBRIDGE UNIVERSITY PRESS
The Edinburgh Building, Cambridge CB2 2RU, UK
40 West 20th Street, New York NY 10011–4211, USA
477 Williamstown Road, Port Melbourne, VIC 3207, Australia
Ruiz de Alarcón 13, 28014 Madrid, Spain
Dock House, The Waterfront, Cape Town 8001, South Africa

http://www.cambridge.org

First published 1995
First paperback edition 2003

A catalogue record for this book is available from the British Library

Library of Congress cataloguing in publication data

Fuller, Robert J.
Bird life of woodland and forest / Robert J. Fuller; with line illustrations by Chris Rose.
    p.   cm. – (Bird life series)
Includes bibliographical references (p.   ) and index.
ISBN 0 521 33118 8 (hardback)
1. Forest birds – British Isles.   2. Forest birds – Ecology – British Isles.   I. Title.
II. Series.
QL690.B65F85   1995
598.252′642′0941 – dc20   94-9647 CIP

ISBN 0 521 33118 8 hardback
ISBN 0 521 54347 9 paperback

To Angela, Ruth, Alice and Edwin

CONTENTS

PREFACE

For as long as I can remember woodland has been a part of my life. I spent my childhood and, as it turned out, nearly the first two decades of my working life, living next to the escarpment of the Chiltern Hills. It was here that my interest in woodland and its birds took hold. Initially there were quests for the more elusive birds of the region including Hawfinches, Redstarts and nesting Buzzards. Later, with a small group of friends, I became temporarily absorbed by birds of prey. Weekends and evenings in the early 1970s were spent tracking down Sparrowhawks. This was at a time when the hawk was in the depths of its pesticide-induced decline. Throughout that period, however, a population persisted in the Chilterns and within a few years we had located some 50 territories. Later still, it was scrub that held my attention. The Chiltern escarpment presented an opportunity on my doorstep that I could not ignore. Much of the downland was rapidly being invaded by scrub, which in some areas was well on its way to becoming woodland. Following earlier work by Ken Williamson, I decided to study the responses of birds to these changes in vegetation.

This theme of habitat change was continued when I studied the effects of woodland management, especially coppicing, on birds. I was fortunate to carry out this work in some remarkable woods throughout Britain and I found myself lured irresistibly by their wider ecology and history. Ancient woods have a potent effect for they are far more than mere collections of trees. The sense of ecological and social history in these places is strong. Some even offer a tantalising link with the prehistoric vegetation of Britain. This awareness of ancient woodland owes much to Oliver Rackham and George Peterken.

If ancient broadleaved woods lie at one extreme of the continuum of British woodland, then the coniferous plantations of the twentieth century are at the other. These new forests increasingly have much to offer the ecologist and birdwatcher but unfortunately most of my own visits to coniferous forests have been essentially casual. In this book I have attempted to write a balanced and contemporary account of the bird life that uses Britain's tree-covered land. In doing so I have relied heavily on the advice and expertise of many other people, but this has been especially necessary in the case of coniferous forests.

The approach of the book reflects my own interest, as an applied ecologist, in trying to understand how man's past, present and future treatment of woodland affects the birds that live in it. In this respect the book differs strikingly from the two earlier books on woodland birds in Britain: *Birds and Woods* (Yapp, 1962) and *Woodland Birds* (Simms, 1971). A second major difference is that I make little attempt to describe bird communities in relation to woods that consist of different tree species. There are no chapters dealing exclusively with the birds of oakwoods, beechwoods and ashwoods for example. Many woods are complex mixtures of different trees and shrubs. Simple botanical classifications must, therefore, inevitably mask important variation or even exclude particularly complex types of woodland. Recent classifications of British woodland acknowledge this complexity (Rodwell, 1991; Peterken, 1993). Purely descriptive accounts of woodland types and their birds are also unsatisfactory because they seldom account for the important role played by woodland structure which may vary independently of the tree species composition.

Strictly speaking, this book is not a guide to the conservation of woodland and its birds, but the subject surfaces repeatedly throughout the pages that follow. I have tried to summarise the opportunities that exist for improving habitats for birds within woods and forests. I hope, therefore, that the book will be useful to those directly involved in the management of woods and forests, though other sources will need to be consulted for detailed management prescriptions. Birds should, however, be kept in perspective – perhaps an unusual view for someone who has worked on birds for the last 20 years! They form just a part of the interesting wildlife to be found in woods and forests and, even in nature reserves, woodland management should seldom be carried out solely with birds in mind.

So many people have helped me to write this book. Thanks to everyone who has discussed ideas, broadened my outlook, provided information, raised my enthusiasm, or offered valuable criticism. I am also grateful to those who introduced me to special places, both in Britain and abroad, and whose company I have enjoyed in them. It has not been easy to pick out individuals, but I would like to take this opportunity to give my special thanks to the following: Per Angelstam, Colin Bibby, Rob Bijlsma, Humphrey Crick, Fred Currie, Paul Donald, Peter Duncan, Pete Fordham, Bernard Frochot, David Gibbons, Andy Gosler, Su Gough, Harry Green, Ted Green, Jeremy

Greenwood, Pekka Helle, Andrew Henderson, David Hill, Shelley Hinsley, Ron Hoblyn, Dick Hornby, David Jardine, Keith Kirby, Derek Langslow, John Marchant, Mikko Mönkkönen, Liz Murray, Ian Newton, Sven Nilsson, Raymond O'Connor, Will Peach, Chris Perrins, George Peterken, Steve Petty, Mike Pienkowski, Carolyne Ray, Pete Robertson, Ken Smith, Jonathan Spencer, Tadeusz Stawarczyk, Juha Tiainen, Ludwik Tomiałojć, Martin Warren, Tomasz Wesołowski and Brian Wood.

Each of the chapters of this book has benefited from a critical reading by several people. For this favour I am extremely grateful to Paul Donald, David Gibbons, Andy Gosler, Ted Green, Richard Gregory, Ron Hoblyn, David Jardine, Keith Kirby, Will Peach, Steve Petty, Robert Prŷs-Jones, Ken Smith, Ron Summers, Bill Sutherland, Ludwik Tomiałojć and Tomasz Wesołowski. Appendix 1 was much improved by advice and information from John Marchant. Chris Perrins not only commented on every chapter but made sure that I actually finished writing the book. Sophie Foulger helped greatly with various stages of production. Thanks also to Tracey Sanderson and Alan Crowden at Cambridge University Press. It has been a pleasure to work with Chris Rose whose drawings and cover painting so effectively bring the birds to life in their woodland habitat. Those who provided photographs and material for other figures are acknowledged individually in the captions. I would also like to thank Alan Knox, Matthew Oates and Bev Walpole for their help in various ways with the figures.

This book was written in a private capacity but it draws to some extent on my own work while employed by the British Trust for Ornithology, and on ideas and contacts I developed as a result of that work. The views expressed here are my own and do not necessarily match those of the BTO, or those of the organisations that provided financial support for my BTO work. These bodies are the former Nature Conservancy Council and the Joint Nature Conservation Committee (the latter on behalf of English Nature, the Countryside Council for Wales, and Scottish Natural Heritage).

Finally, I thank my wife, Angela, for her encouragement and advice throughout the writing of this book.

Robert J. Fuller
Diss, Norfolk
1994

INTRODUCTION

With a mere 10% of its land surface covered by trees, Britain is one of the least wooded countries in Europe. Only Ireland and The Netherlands have less. Barely a fragment of our woodland can be thought of as natural for it has all been influenced by man to a greater or lesser degree. Even those woods that have descended from our primeval forests have been exploited over many centuries to provide grazing as well as the products of the trees themselves. The clearest evidence of man's influence is that a large part of our woodland now consists of plantations of introduced tree species, predominantly conifers. The enormous variety of woodland that now exists in Britain has been created by human activities superimposed on the variations in tree cover that arise naturally from differences in soils and climate. In fact, there is more variation in the woodland and forest cover of Britain than is found in many far larger countries. The central purpose of this book is to describe how birds have responded to the creation of such a wide spectrum of woodland types in Britain and how birds are likely to be affected by ongoing and future changes in woodland. I have pursued this objective through a mixture of review and original observation. The geographical focus is England, Scotland and Wales, though relevant material has been drawn from studies carried out elsewhere.

Relationships between birds and their woodland habitats can be viewed on two levels. The first examines the factors responsible for controlling the abundance and distribution of a particular species. The second level, the community approach, is more concerned with understanding why different numbers and assemblages of species are found in different types of woods. These approaches are complementary and both are adopted here. Many British bird species have been studied in some detail and much is known about the general biology of most – their mating systems, foods, habitat requirements and so on. Some of this information is summarised in this book. However, we have few hard facts on the mechanisms that control the size of woodland bird populations, with three notable exceptions: the Sparrowhawk, Great Tit and Nuthatch. In Britain, the study of bird communities has received scant attention compared with North America and Scandinavia.

Woodland tends to be the most complex of terrestrial habitats, both in terms of its physical structure and in the numbers, and inter-relationships, of plant and animal species associated with it. In trying to understand patterns in the abundance and distribution of woodland birds, it can be extremely difficult to disentangle the effects of one factor from another. Nonetheless, much has been learnt in the last two decades and I have tried to outline current knowledge at the levels of individual species and of bird communities. I have followed this approach in all chapters but some parts of the book are slanted more heavily towards species or to communities. Information on habitats, nest sites, and feeding ecology of all bird species that use British woodland is given in Appendix 1. A summary of the habitats of bird species that use woodland in mainland Europe, but not in Britain, is presented in Appendix 2.

The habitats

The scope of this book is wider than the title might suggest. Essentially, the book is about birds living in *woodland*, which can be broadly defined as any tree-covered land. To qualify as woodland it is not necessary for the trees to be large; coppice and young plantations for instance are covered by the definition. I have chosen to include *forest* in the title to acknowledge the fact that, in Britain, this term has become synonymous with extensive plantations of conifers, though it is also used in the context of our native Scots pine forests. For many people the term *woodland* is applied somewhat uncomfortably to these recent plantations. Nevertheless, for the sake of simplicity, I use *woodland* in the text to embrace both semi-natural woods and plantations. To this core of woodland and forest I have added several other habitats. These include rural parkland, on the grounds that it is an extreme form of wood-pasture, and scrub because it is an early stage of woodland development. Scrub and woodland form a natural continuum of habitats for birds. Lowland heathland is also touched on (Chapter 5) because scrub is an integral, though problematical, element of its vegetation. Scrub invasion is threatening to convert much English heathland into woodland.

The terminology surrounding British woodland is elaborate and potentially confusing. A glossary appears in Appendix 3 and further definitions of silvicultural terms and methods are given in Chapter 1.

The subject of habitat structure crops up time and again in the following pages, so it is desirable to discuss it in some detail at the outset. The *structure* or *physical architecture* of a wood can be broken down into a large number of components, for example tree height, open space, amount of dead wood, volume of the canopy, presence or absence of a shrub layer and so on. From an ornithological viewpoint, the two most important concepts of woodland structure are those of foliage profile (vertical habitat structure) and the openness, or spatial variation, of the wood (horizontal woodland structure).

The foliage profile is the pattern of distribution of foliage from the ground to the crowns of the canopy trees. This is difficult to measure accurately but examples of different profiles are given in Fig. 1, which represents a gradient from relatively simple to complex woodland structures. The foliage within the profile is conventionally divided into several strata. The crowns of the largest trees form the *canopy layer*. Some British woods may have a two-tier canopy but more than

Fig. 1. Three examples of woodland foliage profiles. These increase in structural complexity from top to bottom with the addition of a field layer (centre) and a shrub layer (bottom).

two tiers are rare, although such structures are typical of tropical rain forests. In a lowland deciduous wood, the canopy layer would typically be composed of the leaves and branches of oak, beech, ash or sycamore. Beneath the canopy, the *shrub layer* may consist of small trees and shrubs, such as hawthorn, blackthorn, field maple, hazel and rhododendron. It is sometimes convenient to think of the shrub layer as consisting of the foliage within an arbitrary band of height, usually 1 m to 6 m above ground. Plants growing close to the ground form a further distinct stratum of vegetation, the *field layer*, which includes bramble, ferns, ericaceous shrubs, grasses and herbs. Climbers, notably honeysuckle, ivy, roses and traveller's joy, form an additional component to the vertical complexity of woods.

There are three aspects of horizontal habitat structure (Fig. 2). The first is the density of the canopy, or *canopy cover*, which is essentially the number of gaps in the canopy foliage within stands of trees. Canopy cover is partly a function of the density of the stock and the age of the trees. In plantations, canopy cover is low in the early years after planting it but quickly increases as the trees grow. Canopy cover

Fig. 2. Three aspects of horizontal structure in woodland. *Canopy cover* (top) is shown as the spacing of individual tree crowns with a gradient from a closed stand on the left to a very open stand on the right. *Compartment size* (centre) shows variation in the scale on which stands may be managed within a wood, from large uniform blocks to an intricate mosaic of small patches. *Permanent openings* (bottom) include features such as rides and glades which may vary in frequency from one wood to another.

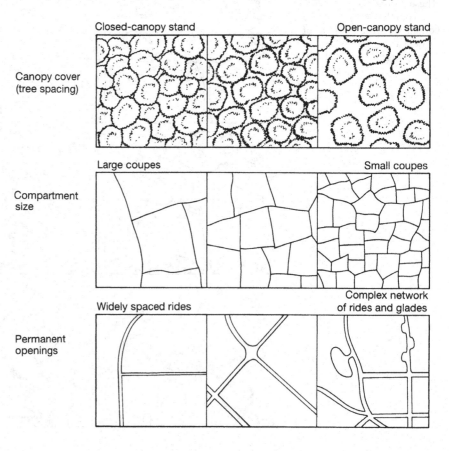

is also influenced by the species of trees because each differs naturally in its foliage characteristics with, for example, spruces casting far heavier shade than larches, and beech more than ash. The second aspect of horizontal structure is the patchiness created by those wood-land management systems that involve trees being harvested within clearly defined compartments on a time cycle. This leads to a patch-work of different stages of growth, which vary in height and foliage profile. At any one point, there will usually be a transition over time from open ground to mature forest, hence these open areas are ephemeral. In undisturbed forest, an analogous patchiness occurs through the natural death of trees by disease, storm or fire. The vegetation structure within these natural treefall gaps is rather differ-ent, however, from that found within a young plantation (Chapter 9). Permanent open areas, such as rides and glades, form the third aspect of horizontal structure.

The birds themselves

Woodland holds more species of breeding birds than any other major class of habitat in Britain (Fuller, 1982). Based on information in Appendix 1, some 113 species will use woodland and scrub in Britain, though 26 of these species only use these habitats incidentally. I estimate that of the 198 species breeding regularly in Britain (defined as those recorded in ten or more British 10km squares in both the *Atlas of Breeding Birds* and the *New Atlas of Breeding Birds* (Sharrock, 1976; Gibbons *et al.*, 1993)), some 49 species (25%) are associated with closed-canopy woodland or scrub, and a further 15 species (8%) use open scrub or young plantations as a typical habitat.

Surprisingly few rare British birds depend on woodland. Some 65 native bird species are thought to have national populations of less than 1000 pairs (Gibbons *et al.*, 1993). Only nine (14%) of these are woodland birds, which I define for these purposes as species which will both breed and feed within closed-canopy woodland and scrub (i.e. the species listed in part A of Appendix 1): Honey Buzzard, Goshawk, Wryneck, Redwing, Firecrest, Crested Tit, Golden Oriole, Scottish Crossbill and Parrot Crossbill. Of 94 species with national populations smaller than 10 000 pairs, just 13 (14%) are woodland birds, the additional four scarce woodland and forest species being Capercaillie, Lesser Spotted Woodpecker, Nightingale and Hawfinch. Important populations of other scarce species are associated with some other habitats covered by this book. For example, Hen Harriers, Short-eared Owls, Nightjars and Woodlarks nesting in young plantations, and Red Kites and Long-eared Owls that nest in closed woodland but mainly feed in open habitats.

Much of woodland bird conservation is concerned with maintaining and, wherever possible, enhancing the richness of bird life in woods. At the present time, there are unprecedented opportunities to do this. Many woodland owners are actively seeking methods of productive management that are compatible with a rich wildlife. In some cases, financial support is available for conservation management in woodland. New woods are being planted at a fast rate, both in the uplands and lowlands (Chapter 1). These new woods offer an opportunity to enhance the variety of wildlife in many landscapes, provided that they are designed and managed sympathetically. If conservation is to be more effectively integrated into productive forestry, it is important to understand exactly how forestry alters the woodland habitats, and how birds and other wildlife respond to these changes. It is also necessary to know what flexibility there is within forestry operations to produce those types of vegetation most beneficial to wildlife. Within woodland nature reserves, demand is also increasing for advice on how best to manage habitats for birds. Reserves provide rather different opportunities for conservation because in most cases commercial production is not the primary objective, although rarely is it completely irrelevant. Whatever the circumstances, coming up with the right prescriptions for management should not rely on guesswork or 'conventional wisdom', but on sound ecological knowledge. There is some way to go before we reach this ideal.

I have pointed out that rather few of Britain's rare species of birds live in woodland, but a valuation that rests on rarity alone would be arid, even contrived. Birds in woodland have a very real cultural and recreational value. Bluebells, primroses and bird song are the essence of the special atmosphere of woodland in spring. The Nightingale has been a unique inspirational source. Richard Mabey (1993) has written 'It is hard to think of another wild creature which has had such a versatile and honoured role. At various times the Nightingale has served as a kind of parish familiar or local wood-spirit; as a symbol, and messenger, of love; as a harbinger of spring and an archetype of natural music.'

1

Britain's woodland environment

This chapter provides a background to Britain's woodland as it stands today. To appreciate how birds respond to woods and their management, one has to start with the trees and the woods themselves.

The twentieth century has been eventful for British woodland. It has seen a transformation as traditional ways of woodland management have been largely superseded by plantation forestry. This was the completion of a process that started with the onset of the Industrial Revolution. A major landmark was reached in 1919 when the Government created the Forestry Commission and gave it the responsibility of implementing a national programme of afforestation. This programme was initially intended purely to create a strategic resource, but over the years it has shifted to embrace economic, social, conservation and landscape functions. The history, economics and operations of British forestry have been covered by Avery & Leslie (1990). They describe how forestry is one of a complex of competing land-uses which, in Britain, has traditionally played second fiddle to agricultural interests. However, with British farming facing an uncertain future, it is likely that more land will continue to be released for tree planting, both in upland and lowland Britain. The long-term need for an efficient and productive home forest industry will probably intensify. Despite vigorous planting, both by the Commission and private landowners, Britain still imports some 90% of its wood and wood-based products, while world shortages of timber are forecast to worsen (Last et al., 1986).

Current woodland distribution

The amounts and types of woodland in Britain today vary enormously from one part of the country to another. The present distribution of woodland (Fig. 1.1a) is a product of two different processes, widely spaced in time. First, the clearance of the wildwood was rather uneven. Some regions, such as the Weald and the Chilterns, appear

Fig. 1.1. The distribution of woodland in Britain. (a) Total woodland as shown by the percentage of land covered by all types of woodland. (b) Broadleaved woodland given as the percentage of the total woodland area. Statistics are given for the counties of England and Wales, and the regions of Scotland, and are drawn from the Forestry Commission's Census of Woodlands and Trees 1979–82 (based on Locke, 1987).

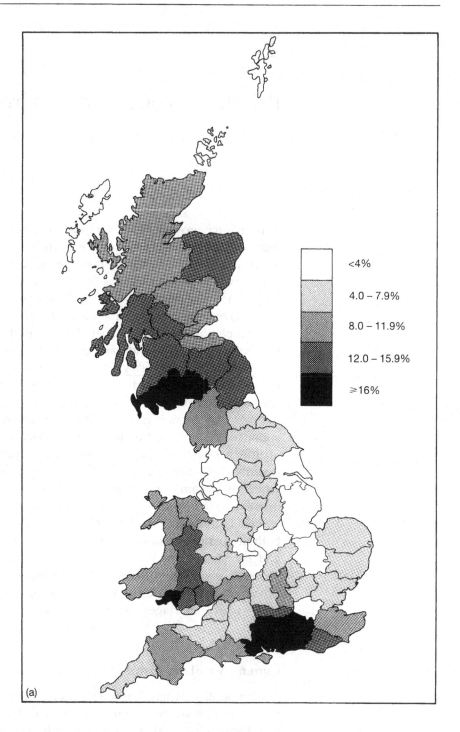

<4%

4.0 – 7.9%

8.0 – 11.9%

12.0 – 15.9%

≥16%

(a)

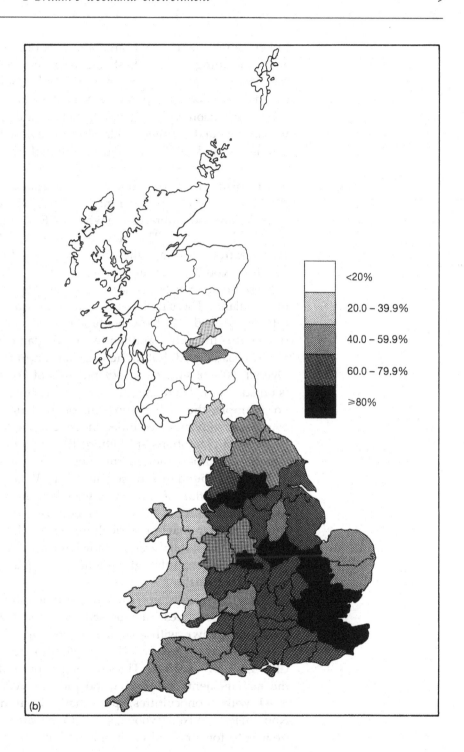

(b)

to have always supported large amounts of woodland, whereas other areas, including much of East Anglia, have remained relatively treeless for hundreds of years. Second, recent decades have seen a huge expansion of forestry, particularly in the north and west.

By comparison with England, both Scotland and Wales are relatively heavily wooded. Within Scotland the heaviest concentration is in the south-west and the border country, though the north-east is also one of the most forested parts of Britain. There is greater regional variation in the distribution of woodland in England than in Scotland and Wales. A large sweep of central England, which extends as far north as Yorkshire and embraces the whole of East Anglia and the Midlands, is sparsely wooded. The most heavily wooded part of England lies in the extreme south, where the most wooded county, Surrey, has just 19% woodland cover (Locke, 1987).

These broad statistics conceal enormous regional differences in types of woodland. The simplest distinction, and one with profound implications for birds, is between broadleaved and coniferous woodland (Fig. 1.1b). In central and south-east England, woodland is predominantly broadleaved, but to the west and north the situation is reversed. Overall, 60% of English woodland, 30% of Welsh and 17% of Scottish is broadleaved. There are, of course, exceptions to this general picture. For example, the high proportions of conifers in Norfolk and Suffolk, compared with other counties in eastern England, are accounted for by the vast plantations at Thetford Forest and along the Suffolk coast. Another example, though this does not emerge strongly from Fig. 1.1b, is the abundance of sessile oak in Wales.

The distribution of coniferous woodland in Britain is a crude mirror image of land productivity. Conifers have been planted on those soils which are marginal for agriculture, especially uplands in the north and west. Substantial tracts of lowland heath and coastal dunes have been planted too, although very little new planting is carried out now in these habitats.

Rather few woods are grown as genuine mixtures, with broadleaved and coniferous trees interspersed. In some regions, broadleaves are commonly grown with a conifer nurse crop which is removed before the broadleaves mature. This applies to beech and cherry in the Chilterns, for example. The challenges presented to the forester in the management of mixed woodland are very different from those faced with monocultures, particularly in thinning and marketing. Nonetheless, mixed woods seem to be attracting growing interest from both forestry and environmental angles (Harris & Harris, 1991).

Recent and future trends in British woodland

The Forestry Commission has made three censuses of British woodland
since the Second World War: 1947–49, 1965, 1979–82 (Locke, 1987).
The overall cover of woodland in Britain rose from 6.7% of the land
area in 1947 to 9.4% by 1980. During this time, there was a fundamental change in the composition of the woodland, with coppice
decreasing from some 10% of the total woodland area to just 2% by
1980, while coniferous high forest, which accounted for less than 30%
in 1947, formed more than 60% by 1980. Changes in the total areas
of the main woodland types are illustrated in Fig. 1.2.

The total area of coppice declined by 73% between 1947 and 1980,
despite a modest revival. In the 1940s, about two-thirds of the coppice
was grown with standards, but by 1980 this had slumped to less than
a third. This reflects the fact that sweet chestnut, which is often
grown without standards, now accounts for about half the total area
of coppice. Over the same time period, broadleaved high forest rose
by nearly 50%; much of this increase stemmed from converted derelict
coppice.

Since the Second World War, the area of coniferous high forest
has increased more than threefold. The shift towards conifers was
particularly rapid between 1947 and 1965, when more than half a

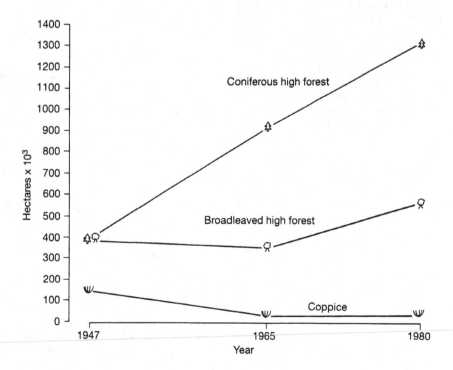

Fig. 1.2. Estimates of the total areas of three woodland types in Britain. Based on three censuses conducted by the Forestry Commission (see Table 46, Locke, 1987). Due to differences in the methods between the three censuses, the trends should be regarded as approximate.

million hectares of conifers were planted. In 1947 and 1965 the dominant coniferous species were, in order of decreasing importance, Scots pine, Sitka spruce, Norway spruce and larch. By 1980, however, Sitka spruce had overtaken Scots pine, and lodgepole pine had emerged as the third most important species, ahead of Norway spruce. In the uplands the switch to planting Sitka spruce has been very strong. At the same time, planting of Scots pine and Norway spruce diminished. For broadleaves, however, there have been fewer changes in the relative importance of species with oak first, followed by beech and ash. The fourth broadleaved species has been birch or sycamore.

During the last decade, planting has continued apace with more than 20 000 ha of new woodland being created in most years. Most of these new forests are coniferous and have been planted mainly in upland Scotland. The rapidity of afforestation north of the border is emphasised in Fig. 1.3, which shows that planting has, by comparison, been negligible in England and Wales.

Away from the uplands, schemes now exist to encourage the planting of new woods on farmland. These woods are mostly small, generally less than 20 ha, but they could do much to increase the variety of habitats available to birds in many areas of lowland farm-

Fig. 1.3. Recent trends in the total area of woodland. Methods of recording changed slightly between 1982 and 1983. The areas of woodland relate to the estimated area of woodland in March each year (Forestry Commission, 1977–1991).

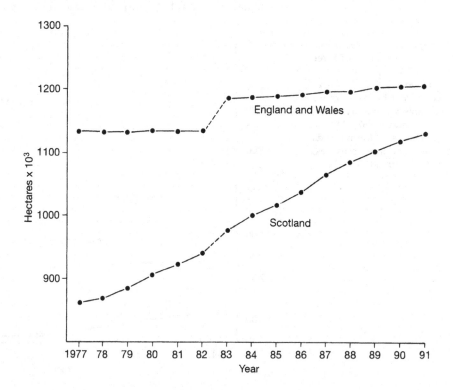

land. Tree planting is increasingly carried out for amenity purposes and in land reclamation (Bradshaw, 1983; Hibberd, 1989). Perhaps the most ambitious project is the Countryside Commission's proposal to develop 12 'community forests' in Britain. The plan is that these will be located at the edges of large urban areas, where there is substantial derelict land, and each will cover at least 10 000 ha. In addition, a new 'National Forest' has been proposed to stretch across a large tract of the Midlands. Time will tell if these new forests materialise on the scale envisaged.

The lowland British landscape differs from that of most other European countries in that it contains many trees that are not located in woods, sometimes presenting an illusion of a well-wooded countryside. Although not strictly the subject of this book, it is worth pointing out that while the total area of woodland has been increasing, the farmed countryside has become more denuded of trees (Peterken & Allison, 1989). At the same time, however, there has probably been an overall increase in the number of non-woodland trees over the last four decades, mainly due to urban and roadside tree-planting, often with non-native species. Hedgerows and their associated trees have been destroyed at a rapid rate over the last 30 years, including many relics of former ancient woodland edges. Since 1969, Dutch elm disease has exacerbated the problem.

Much broadleaved woodland has been converted to conifer plantation since the Second World War. This process was spurred on because so many lowland woods had been devastated in the war effort and the perception was that there was a greater market for fast-growing conifers than for broadleaves. The 1980s saw a change in forestry policy with the introduction of a Broadleaves Policy. Through the availability of substantial grant aid, the objective was to encourage new planting and restocking with broadleaves. Already the policy has stimulated considerable planting of native broadleaves and, it seems, more widespread use of natural regeneration and coppicing (Forestry Commission, 1989a). Conifers will, however, continue to play an important role in lowland forestry, as well as in the uplands, for they are so well suited to modern needs in, for example, the building, pulp and chipwood markets. The relatively fast growth rates of conifers also ensure woodland owners a faster return on their investment.

Conifer plantations are usually composed of alien species, widely considered to be unattractive to a significant part of Britain's native wildlife (this idea may well be true but it has rarely been tested). The canopy of conifer stands often casts such a dense shade that field and shrub layers are poorly developed. Accumulation of conifer

Fig. 1.4. Kielder Forest, located on the English – Scottish border, is one of the largest tracts of spruce plantation in Britain. Planting started in the 1930s. Felling and restocking, mainly with Sitka spruce, commenced in the 1960s. The first generation plantations formed large, even-aged blocks of trees. In future, however, the forest will have more of a patchwork appearance, with coupes of different sizes and carrying different ages of trees. Conifers are no longer planted right up to the streams and these areas are either left open or planted with broadleaves. David Jardine.

needles can lead to increasing acidification of the soil, which in conjunction with the poor light penetration, can create an impoverished ground vegetation. Streams draining afforested catchments have become more acidic, to the detriment of riparian wildlife (reviewed by Gee & Stoner, 1988). Uniform extensive forestry can also be unsympathetic to the relief and character of the landscape, particularly in hill country. In recent years, much progress has been made in tackling these water acidification and landscape issues, with practical guidelines now in place (Forestry Commission, 1988; 1989*b*).

Another change taking place now is that the forests created in the last 50 years are coming to maturity. This gives the opportunity to reshape the forests at restocking, often with wildlife and landscape, as well as forestry, in mind (Fig. 1.4). This restructuring involves looking afresh at the design and layout of the entire forest. At Kielder Forest, for example, vast areas of even-aged Sitka are being fragmented into far smaller felling coupes, and zones of broadleaves are being established along watercourses (Hibberd, 1985; McIntosh, 1989). When fully implemented, this new layout will have several consequences. Visually it will be more in scale with the landscape; the forest will

contain a greater diversity of age classes and habitats per unit area; and the length of edges between different ages of crops will be much longer for a given area. Windthrow is a major limiting factor for upland forestry and one of the aims of restructuring has been to plan the size, shape and harvesting sequence of the new coupes so that the edges are reasonably windfirm.

Forestry has to contend with a variety of insect pests and fungal diseases, neither of which has proved a significant hindrance to the recent expansion of coniferous forestry in Britain. Pines are particularly susceptible to large-scale insect attacks with the most serious, in terms of the area of trees affected, being the defoliation of lodgepole pine in northern Scotland by the pine beauty moth. In an attempt to pre-empt further damage by this moth, large areas of lodgepole pine were sprayed with an organophosphate insecticide, fenitrothion. Such large-scale aerial pesticide applications had not been used previously in British forests and there was considerable concern about their effect on wildlife living within the pine forests. There do not, however, appear to have been any major effects on populations of common songbirds such as Coal Tit (Spray *et al.*, 1987).

The most serious fungal disease of trees in recent times was Dutch elm disease, in which the fungus *Ceratocystis ulmi* was spread by *Scolytus* beetles. Elms started to die in the late 1960s and by the end of the 1980s Britain's elms had been devastated. The consequences for birds were complex (Osborne, 1983; 1985). Few species of birds appeared to decline in response to the death of elms, indeed some, such as tits and woodpeckers, probably benefited from the large numbers of beetles and their larvae available on the dead trees. The widespread felling of dead trees, however, did lead to reductions in the numbers of breeding birds on many farms.

Ancient and semi-natural woods

In England and Wales, ancient woods are usually defined as having existed continuously since at least AD 1600 (Spencer & Kirby, 1992). In Scotland, woods that are thought to have developed naturally before the mid-nineteenth century have been classified as 'probable ancient sites' (Roberts *et al.*, 1992). The ancient status of a wood is not compromised by the felling of its trees, indeed this may have happened many times in its history. An ancient wood cannot be regarded as an undisturbed wood for, historically, it may have undergone many different changes (Willis, 1993). Semi-natural ancient woods are those where the trees and shrubs are native to the site

Fig. 1.5. The Devonshire wood in the centre of this photograph is a microcosm of the recent history of lowland English woodland. The former mosaic of broadleaved woodland and heathland has been largely planted with conifers and a patchwork of compartments at different stages of growth is clearly discernible. A fringe of semi-natural woodland appears to have survived along the northern (lower) boundary. The eastern part of the woodland (left) is an ancient wood known as Paradise Copse. This wood was planted with oak some 200 years ago which was replaced with conifers in recent decades, though oak pollards still fringe Paradise Copse. The western part of the woodland (right), called White Down Copse, was formerly heathland, also planted with oak and more recently converted to conifers. The wood at the top of the frame is Ashclyst Forest which, together with Paradise Copse and White Down Copse, is owned by The National Trust and largely leased to the Forestry Commission. Nature conservation is one of the objectives of the current management in this woodland which includes initiatives to recreate heathland and restore wood-pasture. The photograph was taken in August 1987 by ADAS Aerial Photography, Cambridge. Crown Copyright.

and have not been planted but have arisen from natural regeneration or regrowth from cut stumps.

Ancient woods, especially semi-natural ones, tend to be those of the highest importance to nature conservation. They often hold communities of plants and invertebrates not found in woods of recent origin. These habitats are particularly important for sedentary woodland species which are unable to colonise new sites easily. Most birds, however, are highly mobile and the continuity of woodland cover is of rather little significance in determining the type of bird community that a wood may hold. All other things being equal, there is no reason to suppose that an ancient wood will be different from a recent wood, with respect to the birds it holds. But where ancient woods are managed in different ways from recent woods, or are strikingly different in their tree composition, then this will certainly be of ornithological significance.

The Nature Conservancy Council's inventories of ancient woodland reveal the present distribution and recent changes to ancient woodland (Roberts *et al.*, 1992; Spencer & Kirby, 1992). Ancient semi-natural woodland now makes up approximately 23% of the total woodland area in England, 13% in Wales, and 7% in Scotland. In terms of the land surface area covered, the figures are: England 1.6%, Wales 1.5% and Scotland 1.1%. Ancient semi-natural woodland is distributed extremely unevenly across Britain. Major concentrations are found in the Weald, the New Forest and Hampshire, the Wye Valley, the Chilterns and the Spey Valley. Regions with extremely little include the Borders Region of Scotland, Fife, the north-east Scottish lowlands, the Lancashire plain and Merseyside, Humberside, the Fenland around the Wash, and Breckland. Of ancient woodland existing in England and Wales in the 1930s, some 7% has been cleared, mainly to create more agricultural land. The most extensive changes, however, have been conversion to plantations, usually of conifers (Fig. 1.5). In England and Wales, 38% of the former ancient woodland area is now plantations, while in Scotland the figure is 41%. Overgrazing is also a major problem, which reduces the conservation value of many woods, especially in the uplands. The area of woodland now affected by overgrazing cannot be quantified.

It is possible to discern broad zones in the composition of semi-natural woods in Britain (Fig. 1.6). This is not straightforward because past management has profoundly altered the natural vegetation to a large degree and there are many gradations rather than clear-cut boundaries. Nonetheless, broad patterns are evident that reflect major geological and climatic variation within Britain. Major north–south and west–east gradients in woodland composition are evident from

Fig. 1.6. To place British woods into a wider European context see Jahn (1991).

One woodland zone deserving special mention is the Highland zone (Fig. 1.7), for it supports two of the most distinctive forest types of Europe: semi-natural Scots pine forest and birch woodland. It is extremely difficult to measure the area of these woods, for the boundaries are often indistinct and even change in response to grazing and fire. Some of the pinewoods are extremely open, with trees scattered on open moorland. Of the total area of approximately 10 800 ha of native pinewood estimated in 1971, only 1600 ha was dense woodland (Goodier & Bunce, 1977). The open structure of much of the pine woodland is probably not natural, but rather a consequence of severe grazing pressure. Despite the international importance of this type of woodland, it too has suffered losses over the last 30 years amounting to about a quarter of the surviving relics of the once extensive Caledonian forest. Forestry or felling, followed by heavy grazing, have caused the damage but fortunately most of the remaining fragments are now protected (Bain & Bainbridge, 1988).

Fig. 1.6. Zones in the tree species composition of semi-natural woodland in Britain. Characteristic tree species are indicated in parentheses. In the case of the mixed deciduous zones there are too many characteristic tree species to list them individually, but the south-east zone is distinguished by the presence of beech, hornbeam and sweet chestnut. Based on Kirby & Heap (1984).

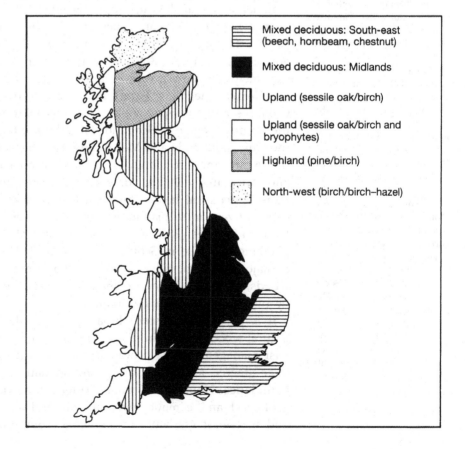

Mixed deciduous: South-east (beech, hornbeam, chestnut)

Mixed deciduous: Midlands

Upland (sessile oak/birch)

Upland (sessile oak/birch and bryophytes)

Highland (pine/birch)

North-west (birch/birch–hazel)

Systems of growing trees

Woodland trees in Britain have been grown and harvested in three basic ways: *high forest*, *wood-pasture* and *coppice*. Each supports rather different bird communities (Chapters 6, 7 and 8), although there is considerable variation within these systems, both in management techniques and in their birds. Coppice and wood-pasture are ancient practices now localised in distribution by comparison with high forest, though coppice is making something of a modest comeback. What follows is the most skeletal outline necessary for understanding the effects of woodland management on birds. For definitive accounts see Evans (1984) and Matthews (1989), while Rackham (1980, 1986, 1990) describes the historical use of our woods.

Most woods in Britain fall readily into one of the above three categories but there are exceptions. The main one is *carr* woodland which is characteristic of permanently waterlogged ground (Fig. 1.8). The dominant species are alder and willows. All forms of scrub lie outside the above categories, as do derelict woods which have not

Fig. 1.7. This view towards Lochnagar, across the valley of the River Dee west of Ballater, Aberdeenshire, epitomises the character of woodland in the Highlands today. Extensive conifer plantations can be seen in the middle distance, with natural regeneration of birch and pine in the foreground. The old Scots pine represents a fragment of the native pine forest. Nick Picozzi.

been managed for many decades. Many of the latter were formerly coppiced.

High forest

Most contemporary productive woodland is managed as high forest. High forest systems produce timber from single-stem trees. Rotation lengths depend on tree species, markets and site conditions, but most conifers are felled at 45 to 60 years of age while some broadleaves, notably beech and oak, are allowed to grow for well over 100 years. High forest includes woodland as diverse as Sitka spruce plantations in Caithness and beech woods in the Chilterns. In Britain, the vast majority of high forest is established by planting, although ash and beech are sometimes regenerated naturally. This is in striking contrast to French woods where there are strong traditions of regenerating all broadleaved trees naturally. Five main systems can be recognised although there are many variants (Fig. 6.4 illustrates four of the systems).

Fig. 1.8. Swampy woodland with a vigorous field layer at Spartum Fen, Oxfordshire. Birds in British carr woodland have hardly been studied, but the densities of breeding birds are likely to be high. Riverine and swampy forests in mainland Europe are extremely rich habitats for breeding birds. Peter Wakely, English Nature.

1 *Clear-felling* The felling and planting of blocks of 0.5 ha to 300 ha. Most conifer plantations are managed by clear-felling. Usually regeneration is achieved by planting. This leads to a rather low variety in horizontal structure.

2 *Group-felling* or *group-selection* Small groups of trees, no larger than 0.5 ha, are felled and regenerated thus creating a patchwork of small stands at different ages. Natural regeneration is commonly, but not exclusively, used. This is becoming more widespread in English broadleaved woods. Group-felling systems are occasionally applied to conifer woods, as in the Bradford-Hutt system (Harris & Kent, 1987).

3 *Selection* Trees are removed individually, with the aim of maintaining a wood of constant appearance with shade-tolerant trees of different ages interspersed throughout the wood. Natural regeneration is used where conditions are favourable. Selection systems are rare in Britain but there is increasing interest in *continuous cover forestry* that embraces single tree selection.

4 *Two-storey system* Underplanting of an existing canopy that has usually been heavily thinned. Where shade-tolerant conifers are planted beneath broadleaves, the purpose is normally to obtain a fast return while retaining the best broadleaved trees. Underplanting is less common than it was 20 years ago when conifers were frequently introduced into broadleaved woods.

5 *Shelterwood* Used locally in Britain, but widespread in French broadleaved woods. It is designed to rely on natural regeneration. When a compartment is felled, a proportion of the trees is left standing as a seed source and to give shelter to the growing seedlings. The seed trees may be left as scattered individuals or in belts. These seed trees are finally removed in one or more fellings when the seedlings are established.

The structure of high forest is extremely variable depending on soil type, management system, tree species and age of the trees. Just as in coppiced woods, there are enormous changes in woodland structure as the trees mature. Generally these changes are more gradual than in coppice. The development of high forest follows a fairly distinct sequence of six stages. *Establishment* is the earliest stage of regeneration. The trees have an extremely open canopy and a dense field layer often develops rapidly. There may be use of inorganic fertilisers, especially in the uplands where poor soils can limit tree growth, and chemical or mechanical weed control. Tree shelters are typically used in small plantations, to protect the trees from deer and other mammals and to enhance growth, but this is expensive for areas larger than about a hectare, which are usually fenced. The *pre-thicket stage* is reached when the trees are some 1–3 m tall. The canopy is still open and the field layer remains dense. Canopy-closure occurs in the *thicket stage* by which time the trees may be up to 10 m tall. Many of the plants on the woodland floor become shaded out and, under conifers (larch is an exception), the ground frequently becomes totally bare. There may be some 'cleaning' or light thinning of the trees at this stage. The main thinning, however, takes place at the *pole stage* when the trees top 10 m. The purpose is to promote the best trees and to ensure that good growth and form is obtained for the final crop. The process of thinning can be beneficial both in forestry and in wildlife conservation terms (Fuller, 1990; Harris & Harris, 1991). In certain upland areas, soils and exposure conditions militate against thinning and this operation is sometimes foregone to allow crops to grow to a greater height and size before the onset of windthrow. When the crop is near to felling age, it is deemed *mature*. Maturity in forestry terms is a very different concept to biological maturity. The forester usually judges it to be the age at which the growth rate begins to slow, or when additional growth is not sufficient to compensate for the loss in revenue accrued from delayed felling. Although the tree may have attained only a fraction of its potential life span, a longer rotation would incur a loss of potential production for that site. Nonetheless, trees are sometimes grown far beyond the commercial rotation for landscape or conservation reasons. Stands of such trees can be regarded as a *retention stage* (other names might be the *veteran* or *over-mature stage*).

Wood-pasture

Wood-pasture is fundamentally different from the other systems in that it combines the growing of trees with the grazing of domestic animals or deer. Although widespread in the Middle Ages, wood-pasture is now virtually defunct in lowland England. However, many of the sessile oakwoods in upland regions of Britain and birchwoods in northern Scotland are similar to wood-pasture, mainly by default rather than design, for many are unfenced. Successful wood-pasture demanded a delicate balance. Too many animals would eventually destroy the wood and this must have happened frequently. Indeed, fears have been expressed for the future of western and northern woods. Many of the largely unwooded commons in the south and east of England had their origins in wood-pasture. A smattering of lowland wood-pasture has survived, for example the Ancient and Ornamental Woods of the New Forest and The Mens in West Sussex.

These relics are famous for their large numbers of ancient trees. Many of these trees were once pollarded by regular cutting, some 3 m above ground beyond the reach of browsing animals, to provide a renewable source of wood (Fig. 1.9). Another striking feature of much wood-pasture is the sparsity of the shrub layer, though holly can grow vigorously as can be seen in many parts of the New Forest.

Parkland is a form of wood-pasture where the balance has tipped towards grazing and the numbers of trees are generally few and scattered. I am not concerned here with urban parks but rather with rural parks, which have two main origins. In the Middle Ages, many parks were created for keeping deer which provided food and sport. These early parks sometimes encompassed substantial tracts of woodland. Rackham (1990) regards the medieval parks as a private form of wood-pasture to distinguish them from Royal Forests and commonland. A few examples survive and virtually all harbour oaks of enormous antiquity. By contrast, most of the rural parks laid out in the eighteenth century were essentially ornamental, although many were created by the landscaping of ancient deer parks.

Fig. 1.9. Ancient wood-pasture with beech pollards, Felbrigg Woods, Norfolk. Peter Wakely, English Nature.

Coppice

Coppicing is the other traditional use of woodland. The distinguishing feature of this ancient practice is that the trees are cut regularly to produce wood of small dimension. The tree grows rapidly from the cut stump (the *stool*) with a characteristic multi-stemmed form. Coppice, and the wood it produces, can be referred to as *underwood*. Most broadleaved trees coppice well. In southern England, the coppiced species include ash, hazel, hornbeam, field maple, alder, birch, small-leaved lime and sweet chestnut. In the north and west, sessile oak was commonly coppiced in the eighteenth and nineteenth centuries. Approximately half the total area of active coppice now consists of sweet chestnut which is common in Essex, Kent and Sussex.

Coppicing was the commonest management system in much of lowland Britain during the Middle Ages but its origins go much further back. Unlike most modern forestry, coppicing does not depend on the planting of trees for it exploits the natural regrowth of shoots from the stools. In its heyday, this renewable woodland resource formed the basis of a large range of products for local communities, including fuel, building materials and all manner of utensils and tools (Rackham, 1980). Coppice also had industrial significance in providing charcoal for iron-smelting and bark for tanning. During the later stages of the Industrial Revolution, these traditional markets and uses for coppice wood collapsed with the wider availability of fossil fuels and an increasing demand for large timber, paper, pulp and particle board. The decline in coppicing has left many woods with much derelict coppice.

The length of underwood rotation depends very much on the tree species and intended market. For example, hazel is cut on a rotation of 7 to 10 years, sweet chestnut about every 15 years, but ash and oak at some 25 to 35 years. The coppice is cut in compartments called, among other names, panels, fells, cants or haggs, depending on the region. In most worked coppiced woods, at least one panel is cut each winter. The size of the panels typically lies between 0.5 and 3 hectares. A 50 ha wood, managed on a 25 year rotation, would require an average of 2 ha to be cut each winter.

Much coppice is grown with scattered single-stem trees known as *standards* (Fig. 1.10). Most are oaks, with ash the second most common. Standards are felled on a much longer rotation than the underwood to provide a modest crop of timber. There should not be too many standards or the underwood will be heavily shaded and growth will be poor (Fuller & Warren, 1993).

Extremely rapid changes occur in the structure of coppiced woodland as the trees grow. These are matched by equally rapid changes in the bird communities (Chapter 6). The ground is usually sparsely vegetated during the first summer after cutting, but the following year there may be magnificent carpets of spring flowers. For at least one further year, the ground cover of grasses and herbs continues to thicken. Gradually, however, many of these light-demanding plants are shaded out as the coppice shrubs grow and eventually form a closed canopy, usually 5 to 8 years after felling. There is a period,

Fig. 1.10. The characteristic structure of coppice with standards, Bradfield Woods National Nature Reserve, Suffolk. This wood is perhaps the most celebrated example of a traditional working coppice. The underwood is predominantly ash, alder, birch and hazel. The standard trees are mainly oak and ash. The coppice in this photograph was probably cut three to four years previously and has not quite closed canopy. Among the characteristic birds breeding in this stage of coppice growth are Dunnock, Willow Warbler and Garden Warbler. Derek Ratcliffe.

Fig. 1.11. The patchwork of coppice panels at different stages of growth is clearly visible in this aerial photograph of Bradfield Woods National Nature Reserve, Suffolk. Individual standard trees can be discerned in some of the panels. This wood has a history of more or less continuous coppice management dating back to the thirteenth century. The western, rectangular part of the woods, known as Monk's Park, is joined to the eastern part, Felshamhall Wood, by a narrow neck of woodland containing an ancient fish pond. In the 1960s about two-thirds of Monk's Park was destroyed to create yet more arable farmland (Rackham 1990). The original boundary of Monks' Park can be easily traced to the south of the surviving woodland. The photograph was taken in October 1985 by ADAS Aerial Photography, Cambridge. Crown Copyright.

typically 2 years either side of canopy-closure, when the low foliage forms an almost impenetrable thicket (Fig. 1.10). This low foliage is, in turn, shaded out and it becomes possible to walk easily beneath the coppice canopy. There are no major changes in the foliage profile of the coppice, although the height and diameter of the stems continue to grow until the next crop of coppice poles is taken.

An actively coppiced wood, where panels are cut every year, consists of a mosaic of patches differing in vegetation structure (Fig. 1.11). Consequently, coppice can offer a wide range of habitats for woodland plants and animals. Partly for this reason, coppicing is widely carried out for conservation reasons. It would be a mistake, however, to think that coppicing favours all woodland species. Coppicing does not generally create suitable habitats for those species requiring dead wood or an abundance of large trees.

2

Historical and European perspectives

The paucity of woodland in Britain is not a recent phenomenon. This, and the long isolation of our woods from those of mainland Europe, have imposed some fundamental restraints on the woodland bird fauna of this island. These can be appreciated only by exploring the historical events that have fashioned Britain's woodland. This chapter sets out to summarise these events and to discuss their significance for birds. In doing so it will become evident that the bird life of British woodland differs in several respects from that of mainland Europe.

The history of Britain's glaciations, the subsequent colonisation by trees and their eventual demise at the hand of man makes an exceedingly complex story here reduced to its minimum essentials. The main sources I have drawn on are Moreau (1954), Yapp (1962), Godwin (1975), Rackham (1986), Harrison (1988), Ratcliffe (1990) and Peterken (1993). Birds do not make good fossils for their bones are weak, so reconstructing past avifaunas is usually a matter of inference based on knowledge of the dominant vegetation and of the habitat needs of the birds themselves. This is risky because a species may change its habitat, or it may show different patterns of habitat selection in different parts of its geographical range. I make this caution because, in discussing past bird communities, I have been forced to make the assumption that birds occupied similar habitats in the past to those they occupy today. Perhaps the best documented example of a bird changing its habitat is the Mistle Thrush in Germany and elsewhere in central Europe (Glutz von Blotzheim & Bauer, 1988). This bird was confined to coniferous forest, but in the mid-1920s it started to spread rapidly, initially into farmland areas, then into suburbs and eventually into urban parks. There are several possible reasons why a species might change its habitat. It may do so in response to a change in numbers of a competitor, or colonisation of the new habitat may be 'overflow' caused by an increase in population size in the initial habitat.

Historical aspects of British woodland and its birds

At the end of the last Ice Age, about 11 000 years ago, much of Britain was treeless tundra. This Ice Age, with one or two respites, had lasted some 50 000 years and at its height few arboreal birds were found north of the Pyrenees. Birds of deciduous woodland had been virtually confined to the southern extremities of the Mediterranean peninsulas, while those of coniferous forest were slightly less restricted. When the climate warmed the trees started their northward 'migration'. The first woody vegetation to establish itself in Britain was scrub composed of birch, juniper, willow and hazel. This developed into taller woodland of birch and hazel, followed by Scots pine. These boreal forests would have resembled those found today in northern Fennoscandia and in the western taiga of Russia, though the British forests contained no spruce. The birds that occurred in the earliest post-glacial woodland would presumably have been similar to those found today in these northern continental regions. Common breeding birds in the birch may have included Willow and Black Grouse, Redpoll, Willow Warbler, Meadow Pipit, Tree Pipit, Fieldfare, Redwing, Brambling, Bluethroat and Waxwing; with waders, like Greenshank, Wood Sandpiper and Spotted Redshank, on the forest bogs and swamps. The pine forests were the habitat of Capercaillie, Siskin and Crossbill. It is not known whether boreal species such as Great Grey Owl, Ural Owl, Hawk Owl, Siberian Tit, Rustic Bunting and Arctic Warbler, ever spread as far west as Britain. On the other hand, the bones of Hazel Grouse and Pine Grosbeak have been found; both species probably became extinct as the climate gradually warmed.

Other trees gradually followed – alder, oak, lime, elm, maple – and with them a greater number of bird species. By the time that Britain became separated from the continent, approximately 7800 years ago, the only extensive unwooded areas were probably saltmarshes, dunes, fenland, the tops of high mountains and parts of the flow country in the far north of Scotland. Most of this great natural forest, which Oliver Rackham (1990) has called the *wildwood*, was deciduous. By this time pine had probably moved north to occupy its present natural distribution in what is now Scotland. Many species of plants and animals present on the mainland had simply not spread into Britain by the time the English Channel was created. To this day, English woodland lacks species of plants and animals that are to be found in similar habitats in nearby mainland Europe. In the case of woodland birds there are some 30 species that are found in the woods of mainland Europe but not in Britain (Appendix 2). By contrast, there are no species that are confined to British woodland. Some of these 'missing' species may have reached Britain only to vanish in the face of the subsequent massive forest destruction, which I describe later. It seems that a high proportion of the birds that failed to arrive, or at least to establish themselves in Britain, were woodland species (Harrison, 1988). As far as we know, Black Kite, Black Woodpecker, Middle Spotted Woodpecker, Grey-headed Woodpecker, Icterine Warbler and Melodious Warbler never did colonise. Pygmy and Tengmalm's Owls are unlikely to have established themselves in Britain because they are associated with spruce, which was absent as a native tree during the post-glacial period.

While Britain is relatively poor in woodland bird species compared with mainland Europe, Ireland is even more impoverished. In Ireland there are no Tawny Owls, woodpeckers, Marsh Tits, Willow Tits or Nuthatches. Some other woodland species are extremely rare in Ireland, including Tree Pipit, Redstart, Wood Warbler and Pied Flycatcher. It is not known if any of these species were once established there, though sub-fossil remains of Great Spotted Woodpecker have been found (Hutchinson, 1989).

The 2500 years following the breach of the land bridge were relatively stable ones for British woodland until the clearances by Neolithic Man. By 4000 years ago, there were substantial tracts of open country, for example in Breckland and on the southern chalklands. The clearance intensified during the Bronze and Iron Ages and by 1500 years ago it was probably largely complete. By the time of the Domesday Book, England was sparsely wooded. The last two centuries have seen further clearances in two main phases, 1840 to 1870 and 1950 to 1980; these left British woodland in its most impoverished state for 8000 years. But the history has not been one of unremitting destruction. Not quite. During the Dark Ages much farmland was abandoned, falling back to secondary woodland through natural succession. The afforestation of the twentieth century, described in the last chapter, has given some regions a far higher woodland area than they have had for many hundreds of years, although the new woods in no way resemble the old. The pattern of these historical changes is summarised for one region of lowland England in Fig. 2.1.

What of the effects of the destruction of the wildwood on birds? For arboreal birds these must have been nearly as extreme as those wrought by the Ice Ages. Many woodland birds must have suffered huge population declines as their habitat shrank to a small fraction of its former extent. Some authors (Wilson, 1977; Tomiałojć & Wesołowski, 1990) have argued that the impoverished woodland bird communities of Britain and Ireland are largely attributable to the fact that these islands are now among the least wooded parts of Europe.

Hutchinson (1989), however, thinks it is unlikely that many woodland species in Ireland disappeared as a result of forest clearance. It is impossible to pinpoint species that became extinct as a result of this huge historical loss of habitat. Part of the difficulty lies in knowing whether birds such as Black Woodpecker were ever present. Britain and Ireland would always have been at the extreme fringes of the range for many species, not simply geographically but climatically. We cannot be certain whether the climate and habitats (our native forests were poorer in tree species than those of Europe) would have ever been suitable for the missing species and whether viable populations could have ever become established. The exact reasons for the low numbers of bird species in British woodland make an intriguing point for speculation but it seems likely that the early isolation of Britain, climate and forest destruction have all played a part.

Likely candidates for extinction as a result of forest destruction would have been species that needed stands of old timber, such as woodpeckers, or those confined to vast tracts of forest where there is minimal disturbance. Some large birds of prey are potentially vulnerable in this way. Eagle Owl once lived in Britain and was possibly a victim of forest destruction. Honey Buzzard may once have been more widespread, though in Britain it could be limited by climate as much as by habitat. Extinctions of birds of prey could, however, have been as much a result of human persecution (which admittedly may have become easier as the forests were fragmented) as of the devastation of the forest itself.

The effects of habitat loss are more complex than a simple arithmetic decrease in the amount of habitat available for birds. Other factors,

Fig. 2.1. A summary of changes in the amount of woodland (upper) and the major agents of change (lower) in Rockingham Forest, Northamptonshire, over the last 5000 years. From Peterken (1993) *Woodland Conservation and Management*, 2nd edn. published by Chapman and Hall.

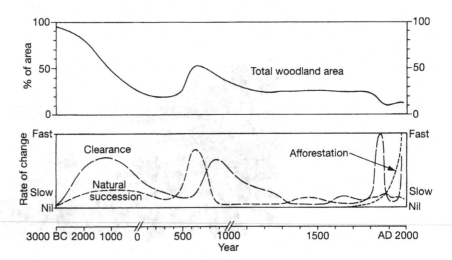

such as the ability of a species to move between habitat fragments, as well as the sizes of the fragments themselves, have a bearing on the likelihood of populations surviving. Patterns of predation would change in complex ways. The clearance of natural forest is likely to lead to an overall reduction in predators, with a consequent long-term increase in breeding success as argued by Wesołowski (1983). On the other hand, forest destruction can have the effect of exposing birds to predators, such as crows, which are associated mainly with open countryside. Nest losses for ground-nesting birds have been found to be highest at the edges of woodland (Andrén & Angelstam, 1988). This can lead to a situation where only the largest patches of forest are suitable for some species. For every species that lost out to the destruction of the wildwood there was at least one that gained. Birds like Buzzard and Rook that nest in trees but feed over open country benefited. Birds of the forest edge would also have increased at the expense of species that avoided the edges and only lived in large blocks of forest.

This historical story of forest decline was repeated throughout much

Fig. 2.2. Displaying Honey Buzzards are a rare sight in Britain, but this elusive species is characteristic of mature woodland throughout much of mainland Europe. It is possible that the climate of Britain is too damp and cool for this summer visitor. Chris Rose.

of lowland western Europe, though far more native woodland was allowed to survive in France, Belgium and Germany. The early Britons seem to have done a particularly efficient job in clearing the land. In the far north of Europe, people started to make really serious impacts on the boreal conifer forests only during the present century. Here the consequences for the birds are, therefore, far better understood and they make an interesting comparison with what is thought to have happened to woodland bird communities in Britain several thousand years earlier. In Fennoscandia, the pattern of human land-use has not involved conversion of forested land to farmland but more intense exploitation of the forest itself. Over the last 40 years, the area of old natural forest has declined in Finland and Sweden as commercial harvesting has assumed greater economic importance. Even in the far north this has had three impacts on the structure of the forest. First, the forest has been fragmented by clear-cutting into sharply defined patches carrying different ages of trees, so the total length of edges has increased. Secondly, the age structure of the forest has shifted towards a predominance of young and middle-aged stands. Thirdly, old trees and dead wood are now more or less absent from large areas.

The effects on bird populations have been enormous and these have been recorded especially thoroughly in Finland (Helle & Järvinen, 1986; Väisänen *et al.*, 1986; Virkkala, 1987; Haila & Järvinen, 1990). Those birds that depend on the old stands of boreal forest have gone into decline; among these species are Capercaillie, Three-toed Woodpecker, Siberian Tit, Siberian Jay and Redstart. Detailed studies of the Siberian Tit have shown why much of the northern forest has deteriorated for this species (Virkkala, 1990; Virkkala & Liehu, 1990). The birds avoid bushy vegetation, preferring areas with dead trees, birches and large conifers. Dead and large trees are removed in thinning operations, which leads to a reduction in the breeding success of the tits. Among those species that appear to have benefited from the changes in forestry are those that use shrubby habitats such as young forest. Examples are Willow Warbler, Garden Warbler, Song Thrush and Robin. In conservation terms, these are widespread, successful 'generalist' species and increases in their numbers cannot be viewed as compensation for the losses among the old forest specialists.

The wildwood and its birds

The composition of the natural forest has been worked out mainly by studying deposits of pollen laid down in such places as peat and the muddy bottoms of lakes (Huntley & Birks, 1983). Perhaps the most unexpected feature is that in lowland England the commonest trees were small-leaved lime, hazel, oak and elm. Though very local in England today, lime was probably the dominant species. Rackham (1986) has divided the wildwood into five provinces: *lime* in south-east Britain, *hazel–elm* in south-west Wales and much of Ireland, *oak–hazel* over much of northern and western Britain and western Ireland, *pine* in central Scotland, and *birch* in the far north of Scotland.

The structure of this forest, at least in the lowlands, would have been rather different from extant woods. This can be judged from largely natural stands that survive in other temperate countries. The most extensive are in North America but fragments do remain in Europe (Fig. 2.3). Eastern Poland holds the largest relic (47 km^2) of undisturbed forest in Europe in the Białowieża National Park (pronounced bee-owa-veea-zha). To say that this tract is entirely natural would be misleading, but it may never have been managed systematically, and can certainly be regarded as 'near-natural' or primeval forest. There are similar, smaller fragments of apparently natural forest elsewhere in eastern Europe, especially in the former Czechoslovakia, but in western Europe there is nothing remotely comparable (Peterken, 1992). Perhaps the human activity of greatest significance in Białowieża has been the stocking of game animals at artificially high levels, particularly during the nineteenth century when it was a sporting ground of Russian tsars. Nonetheless, it remains one of the very few places where it is possible to see the structure of natural forest and consequently to assess how man has altered our forests and their bird life through management. A mixture of hornbeam, lime and oak covers nearly 50% of the area of the National Park. The remainder is taken up with swampy alder–ash forest and coniferous forest. The birds of these habitats have been recorded in great detail (Tomiałojć *et al.*, 1984; Tomiałojć & Wesołowski, 1990). When I worked there in 1988, I was particularly struck by seven features of the structure of this Polish forest.

1 The sheer height of the trees: oaks, limes, ashes and spruces commonly exceed 35 m, although rarely do their trunks measure much over a metre in diameter.
2 There were huge amounts of dead wood, both standing and fallen.
3 Unlike so much lowland English woodland, there was little field or shrub layer, except in swampier areas and in treefall gaps with dense regeneration.
4 The trees formed several layers of canopy.
5 There were many treefall gaps, ranging in size from a single tree to one area about 200 m by 800 m. The larger treefall gaps were

created by storms. There were no natural permanently open areas.

6 Some patches of forest were very uniform in their age structure, almost certainly as a result of natural regeneration following a past catastrophe that created a treefall.

7 Extensive areas of this forest were permanently swampy or periodically flooded.

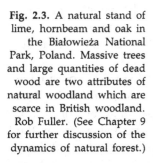

Fig. 2.3. A natural stand of lime, hornbeam and oak in the Białowieża National Park, Poland. Massive trees and large quantities of dead wood are two attributes of natural woodland which are scarce in British woodland. Rob Fuller. (See Chapter 9 for further discussion of the dynamics of natural forest.)

Armed with this knowledge of the structure of natural forests, albeit in eastern Europe, we can make some informed guesses about the nature of the breeding bird communities that used the primeval forest in lowland England. Before doing so, however, I stress that this exercise is fraught with speculative danger. A species may change its habitat with the passage of time or, as I discuss later, may use different habitats in different parts of its range. The climate and soils of eastern Europe are not the same as those in western Europe, so it is also possible that the structure of natural forest, as well as its tree species, may differ between the two regions.

Over a large part of the country the commonest species would have been those that nested on the ground, in cavities, or in the tree canopy. Species that need a well developed understorey, or scrub, may have been confined to the damper areas, treefall gaps, or the edges of the forest where it petered out into open wetland, sand dune or upland. This is the case in Białowieża, where warblers such as Chiffchaffs, Blackcaps and Garden Warblers are found mainly in the swampy forest, which has a dense undergrowth, or in treefall gaps (see also Chapter 9). Some species, such as Willow Warbler and Thrush Nightingale, are surprisingly scarce in the forest, rarely to be found even in treefall gaps. Some of our most familiar woodland birds may, therefore, have been distributed very sparsely in the wildwood. Species that sometimes nest in woodland but feed mainly in open country would also have been uncommon, for example Crow, Rook and Magpie. Birds that depend on holes or dead wood, such as woodpeckers and flycatchers, may however have been more abundant in the wildwood than in most contemporary woods. The dry broad-leaved stands in Białowieża are rich in Wood Warblers and flycatchers. In modern British woods, bird communities rich in Wood Warblers and Pied Flycatchers are confined mainly to upland oakwoods in western and northern Britain (Fuller & Crick, 1992). It is possible that these species were common and widespread in the wildwood.

The Polish primeval forest is rich in breeding bird species. Tomia-łojć & Wesołowski (1990) give a total of 107 species for Białowieża. This exceeds the total number of species for which woodland and scrub are major habitats in Britain (Appendix 1). Many of the species in the Polish forest occur in low numbers, which is similar to the situation in some tropical forests. More than 50% of the species in Białowieża occur at densities below three breeding pairs per km^2. Typically, a far lower proportion of the species in British woodland occurs at such low densities. As an illustration, consider the densities of birds breeding in 88 ha of coppice and oak high forest in Buckinghamshire (R.J. Fuller unpublished). Taking a 7 year period in the

1980s, just 15 out of 51 breeding species (29%) occurred at densities of less than three pairs per km².

The overall densities of birds (in breeding pairs per 10 ha) in Białowieża are of the order 70 to 100 in the riverine forests, 60 to 80 in the lime–hornbeam and approximately 40 in the conifers. The densities of birds in the broadleaved habitats of Białowieża are broadly comparable with those in many large British lowland woods. Three examples, each from woods of at least 40 ha censused in 1987, are: Bradfield Woods, Suffolk, an actively coppiced wood – 83 pairs per 10 ha; Rushbeds Wood, Buckinghamshire, mainly derelict coppice – 53 pairs per 10 ha; Sheephouse Wood, Buckinghamshire, oak high forest – 72 pairs per 10 ha.

Although overall densities may be similar, there is no doubt that several species in Białowieża live at far lower numbers than they do in many British woods. Good examples are Wren, Robin and Blackbird. To some extent this may arise from differences in habitat structure, perhaps the sparsity of the shrub layer in the primeval forest. Furthermore, many English woods include relatively large areas of woodland-edge habitat that may act to elevate their densities (Chapter 4). However, other factors may be operating. The British climate may suit some species, perhaps enabling them to rear more young. Another possibility is that these low densities of birds could be linked to the high rates of predation found in the primeval forest. A large range of predators survives there, both birds and mammals. It is also possible that competition between species plays a role. In the absence of competing species, some birds may be able to live at much higher densities in British woods. If climate or competition are important factors, then it is possible that the British wildwood may have held appreciably higher densities of many songbirds than is the case in Białowieża. If, however, predation is the key factor, then densities in the wildwood would presumably have been much lower than they are in many modern woods.

Geographical differences in habitat selection

On his first visit to a wood, a good naturalist is able to make a list of the birds likely to live there simply by looking at the vegetation. This is only possible because many species have apparently well-defined requirements – some prefer coniferous, some deciduous woods; some need a dense shrub layer, others avoid such areas; some need mature trees and so on. In the chapters that follow, many of these preferences, and their implications for the types of bird communities found in Britain's woodland, will become more evident. It is an interesting fact, however, that not all species consistently select the same habitat throughout their range. A knowledge of the habitats of birds in Britain will not necessarily stand you in good stead on mainland Europe. The Treecreeper lives in many different woodland types in Britain but elsewhere in western Europe its main habitat is coniferous and mixed woodland. The Firecrest, a scarce bird breeding mainly in lowland spruce in England, lives typically in broadleaved woodland in Europe. Willow Tit is a bird of coniferous forest in Fennoscandia, but in Britain it is associated with broadleaves. In southern Europe, several species including Great Spotted Woodpecker and Robin, are mainly found in mountain forests. There are also differences between Ireland and Britain. Bird censuses carried out in 1973 showed that the Goldcrest was one of the commonest breeding species in the Killarney oakwoods, which is in complete contrast to oakwoods in England and Scotland (Batten, 1976). These geographical differences in habitat use are summarised in Appendix 1.

The differences between western Europe and eastern Europe are especially striking. Coal Tit, Goldcrest, Bullfinch and Mistle Thrush in eastern Poland are restricted to coniferous forest, but in Britain they occupy a wider range of habitat. In eastern Poland, Stock Dove breeds more commonly in coniferous than deciduous stands, which is not the case in Britain. In the east, Wood Warbler is very common in mature coniferous forest but it is almost exclusively a bird of broadleaved woodland in the west. Woodpigeon, Blackbird, Robin, Wren and Dunnock are rarely found outside large tracts of forest in eastern Poland. Other more subtle geographical differences of habitat probably await discovery. An elegant example is that of the Black-throated Green Warbler, which is associated with different structures of forest vegetation in different parts of North America (Collins, 1983).

The case of the Mistle Thrush in Germany, described at the beginning of this chapter, shows how quickly a species can expand its suite of habitats. Why some species have done this only in parts of their range is not known with any certainty. Where forest destruction reached its zenith, many species may have avoided extinction only by broadening the range of habitats they used. In very general terms, the trend among many species towards occupying a wider range of habitats in western, rather than in eastern, Europe is consistent with this 'ecological bottleneck' hypothesis. There could, of course, be other causes of geographical variation in habitat use. The presence of a dominant competitor may restrict the number of habitats a species can use. This possibly occurs in the Treecreeper, since in lowland European broadleaved woods its niche would appear to be taken by the Short-toed Treecreeper.

It must also be borne in mind that different regions may offer very different types and mixtures of habitats, which may lead to contrasting patterns of habitat selection that are more apparent than real. The Hazel Grouse is a case in point. This bird needs spruce and alder for cover and for winter food respectively (Swenson, 1993). In Sweden, spruce is far more abundant than alder, but in Poland the reverse applies. The distribution and amount of alder, especially of tall trees growing in close proximity to spruce, is a good predictor of the occurrence and numbers of Hazel Grouse in Sweden (Swenson, 1993). In Poland, however, the species is more obviously associated with spruce patches, but alder appears to be less important, simply because

Fig. 2.4. The vertical root-plate of a fallen tree is a favoured nest-site of the Wren in primeval forest in eastern Poland. In this part of eastern Europe, the Wren is more or less confined to large tracts of forest. This is in complete contrast to Britain where the bird is widely distributed in woodland, scrub, hedgerows and gardens. Chris Rose.

it is so common that it does not limit the distribution of the bird (T. Wesołowski personal communication).

The habitats selected by a bird may vary both in time and in space. It is as well to remember that many birds in Britain may have developed their current habitat use relatively recently and that these patterns may be atypical of a large part of their range.

3

How birds use woodland

This chapter outlines the basic characteristics of Britain's woodland bird life. It is also about the ways in which birds live in woods throughout the year and the factors that shape these lifestyles. Details of habitat use for individual species can be found in Appendix 1. It seems logical that this account should follow the annual cycle for there are large seasonal differences in the uses made of woods by birds. The reader will soon realise that far more is known about woodland birds during the breeding season than any other time of year. But first, some words about population and community concepts that are fundamental to understanding the ecology of woodland birds.

Populations and communities: a woodland view

Birds live in *populations* consisting of individuals of the same species. The bird life of a wood consists of several different populations which form a *community*. Both populations and communities of birds differ greatly from one wood to another, and according to the season, in response to many different factors that are discussed here and in the following chapter.

Changes in the numbers of birds within a population are driven by two main types of processes. The first influence populations independently of their size. Weather and food supplies in winter are two main *density-independent* factors affecting numbers of woodland birds. For many resident species, winter is the main limiting season and spring densities can often be predicted from conditions during the previous winter. Decreases in breeding numbers following a severe winter are most striking among the smaller insectivorous birds including Wren, Dunnock, Robin, Goldcrest, Long-tailed Tit and Coal Tit. Snow cover, rather than frost or low temperatures, appears to be particularly important to overwinter mortality (Greenwood & Baillie, 1991). In some woodland birds, breeding densities are also influenced by the amount of food available in winter. There is evidence that this is the case in the Nuthatch (Nilsson, 1987) and Great Tit (Perrins, 1979), where numbers of breeding birds tend to be higher in springs that follow a good crop of beechmast.

Density-dependent factors, by contrast, are those which have an increasingly severe effect as the population density rises. Such feedback should, in theory, promote long-term stability in numbers and such effects are generally thought of as *regulatory*. Regulatory processes cause the population to return to approximately its former level after occasional reductions caused by factors such as bad weather or poor food supply. It is now widely accepted that the numbers of many birds are regulated in some way. Among British woodland bird populations there is considerable evidence of density-dependent regulation (Greenwood & Baillie, 1991). In most species, however, the exact nature of the regulating factors is poorly understood. One reason is that extremely long-term studies are needed to tease out subtle or intermittent density-dependent effects. Regulation can operate through four different avenues: by inhibiting recruitment to the breeding population; by reducing breeding production; through reduced autumn or winter survival; or, through increased emigration. The fairly sparse evidence to date suggests that density-dependence in birds can operate both within and outside the breeding season but that factors operating in autumn or winter are generally the most important in the population dynamics of woodland birds. A high proportion of population studies on birds have reported density-dependent mortality of juveniles in their first autumn or winter (Sinclair, 1989). In the Great Tit, for example, it appears that recruitment into the breeding population is determined by density-dependent winter mortality (McCleery & Perrins, 1985). In other words, fewer birds survive the winter and take up territories when the population level is high. In the Nuthatch, autumn is the time when juveniles attempt to establish a territory – if successful they may occupy a territory for life. Autumn recruitment of juvenile Nuthatches is lower when adult numbers are high, so the population appears to be regulated by events before the onset of winter (Nilsson, 1987). Hence, autumn and winter are critical times in the life cycle of many resident woodland birds, for the main changes in their numbers appear to be driven by events at this time of year. This also appears to be the case for migrant species such as Blackcap, Whitethroat and Willow Warbler (Baillie & Peach, 1992).

Over the years, much effort has been expended trying to understand how communities of animals are constructed. Much of this work has focused on bird communities, and woodland has been examined in considerable detail because it is structurally complex and rich in species. Despite this, we still know rather little about whether there are any underlying principles determining how many and what sorts of species can coexist in a particular habitat. There are truly daunting problems to overcome in really understanding what makes bird communities the way they are (Wiens, 1989).

The traditional view has been that no two species can occupy precisely the same niche. This idea is widely known as the *competitive exclusion principle*. More recently, the principle has been developed into elaborate theories concerning the numbers of species that can be packed into a community, taking account of the similarity of their niches. It is undeniable that many closely related species show detailed differences in the ways in which they use their environments, either with respect to their habitats, nest sites, feeding sites, or foods and that they often exhibit what are generally taken to be physical adaptations to their preferred niche (Lack, 1971). One of the classic examples is the European tits, which are separated partly by habitat and partly by feeding site. Crested and Coal Tits are 'conifer species', Blue Tit and Marsh Tit are 'broadleaved species', while Great Tit and Willow Tit occur in both conifers and broadleaves, but in Britain their densities are far higher in the latter. The bills of the conifer species are narrower relative to their length, which is presumably an adaptation for probing for insects among conifer needles. The tits also show complex differences in the parts of the tree in which they forage and the heights above ground at which they feed. Such differences, amounting to what is often called *ecological isolation*, are thought to have evolved to enable potentially competing species to coexist in the same habitat.

In the 1960s and 1970s, a huge amount of work was based on the premise that competition between species was the overwhelming force shaping communities. Competition was usually assumed to be for food, though it could also be for other resources potentially in short supply, such as nest sites or roost sites. This was generally accompanied by the notion that communities were in a state of equilibrium; in other words, for any habitat the numbers and types of individuals and species were somehow predetermined and remained stable over time. These ideas have been increasingly called into question and some ecologists believe the most appropriate concepts are simply those of collections or assemblages of species, each with their own independent requirements and interacting to a limited degree. Reality probably lies somewhere between these two extremes.

A real problem is the difficulty of demonstrating the existence of competition and its effects in nature. In general, one would expect the most closely related, or ecologically most similar species, to be in the greatest competition because they tend to use similar resources.

There have been a number of studies that have examined the foraging niches of species in the presence and absence of potential competitors. If competition occurs, then some change in niche use would be expected. Several approaches have been taken. Islands can offer natural experiments in this respect. One example comes from the Swedish island of Gotland which supports just one tit, the Coal Tit, unlike mainland Swedish conifer forests which also have Crested, Willow and Marsh Tits (Alerstam *et al.*, 1974). On Gotland, the Coal Tit uses a wider range of feeding sites than it does in the mainland forests, including several of those typically used by the other tits. In the previous chapter, analogous cases of possible niche expansion or ecological release were recognised for woodland birds in Britain. Better still is the evidence from removal experiments. When a dominant species is removed from its habitat, one would predict that a competitively inferior species might move into the vacated space if it is suitable. In this way it has been shown in Oxfordshire that Blackcaps can exclude Garden Warblers from some otherwise acceptable breeding territories (Garcia, 1983). On islands in western Scotland, Great Tits and Chaffinches hold mutually exclusive territories, though in adjacent

Fig. 3.1. Research in Fennoscandia has shown that the use of winter feeding sites by Goldcrests is influenced by competition with tits. The Goldcrest is one of the few European songbirds that nests in the woodland canopy. In North American forests a considerably higher proportion of bird species are canopy-nesters. Chris Rose.

mainland woods they overlap (Reed, 1982). Removal of Chaffinches from the islands was followed by Great Tits occupying the vacant space, strongly suggesting that the two species were in direct competition. Coexistence of the species on the mainland may have been possible because this was a richer environment than the islands.

Some of the most conclusive work on competition has been on tits and Goldcrests in Fennoscandian conifer forests (reviewed by Dhondt, 1989). Observations and experiments involving the removal of birds have demonstrated that some species are effectively kept out of certain foraging sites by other species that are either behaviourally dominant or simply better at exploiting the food on offer. In winter, Willow Tits avoid the tree species or those parts of trees used by Crested Tits and Great Tits. Also in winter, Coal Tits and Goldcrests feed mainly in the outer parts of trees (i.e. in the external foliage) in the presence of the larger Willow and Crested Tits. Removal of the two larger species led to the two smaller species making far more use of the inner parts of trees.

The evidence is strong that competition can influence the ways in which *some* birds use *some* of their habitats. It would, however, be misleading to imply that all the patterns of habitat use shown by woodland birds are underpinned by competition between bird species. Birds sometimes compete with other animals. An example is provided by wood ants, which can deplete insect populations on trees close to their colonies. A Swedish study found that insectivorous birds such as warblers, tits and Treecreepers visited these trees less frequently than trees that had no ants (Haemig, 1992). There is increasing evidence (e.g. Ekman, 1986; Kelly, 1993; Suhonen, 1993) that predation pressure can be an important factor in determining selection of habitat, nest site or feeding site. Other factors may also have operated in the evolutionary history of a species, leading to patterns of habitat use that have nothing to do with contemporary competition.

Some attributes of breeding communities

Woodland bird communities are made up of species that differ hugely in their numbers and distribution. Some are widespread and abundant, others widespread but less numerous, others are confined to relatively few woods. This is well illustrated in Fig. 3.2, which projects a kind of 'community panorama' for breeding birds in British woodland. It is based on information drawn from 240 woods, of many different types, throughout Britain. In general, those species that are numerically the most abundant ones within woodland are also very widespread, that is, they occur in the greatest proportion of woods. Eight

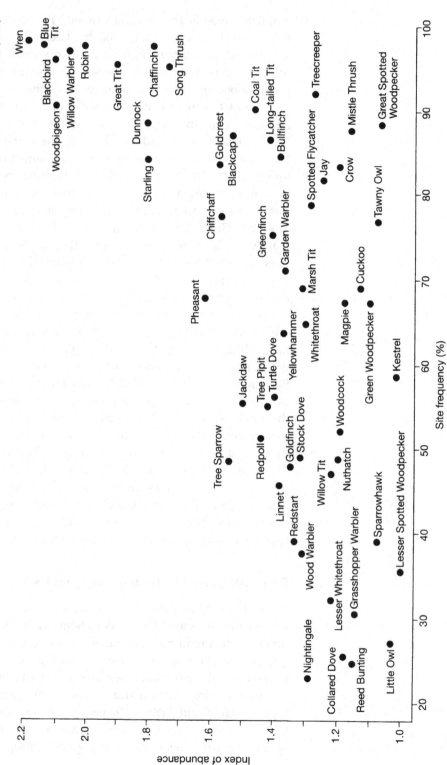

Fig. 3.2. The relationship between the abundance of breeding woodland birds and the frequency with which they occurred in a sample of 240 British woods. The abundance index measures mean abundance at woods occupied by the species. Site frequency is the percentage of woods at which each species was recorded. From Fuller (1982).

species are virtually ubiquitous: Wren, Blue Tit, Blackbird, Willow Warbler, Robin, Great Tit, Chaffinch and Song Thrush. Together with Woodpigeon, Starling and Dunnock they form a core of 11 particularly common species in British woodland. All are residents, except Willow Warbler. These birds are ones that have adapted well to the wide range of woodland habitats now found in Britain. It should be noted though, that these data were collected in the 1970s. Since then British Starlings have declined and the bird has ceased to breed in many woods. None of these core species is confined to woodland. With the possible exception of Willow Warbler, they are among the commonest birds occurring in farmland hedgerows and gardens.

The natural densities of bird species differ considerably. Body size and diet are two factors broadly related to density. Larger species tend to have larger territories and lower densities, though Woodpigeon is a striking exception, perhaps related to its exceptionally varied herbivorous diet (see below). Scavengers and birds that feed on vertebrates are generally much scarcer than insectivores and seed-eaters. These broad patterns can be seen in Fig. 3.2. For example, Crow, Mistle Thrush and Great Spotted Woodpecker are widespread species but they typically occur in much lower numbers than most of the small passerines that are equally widespread. All of the raptors and owls, and to a lesser extent corvids (the crow family), occur at low densities. The most widespread species include a far higher proportion of passerines than the more localised species. Of the species recorded in more than 75% of the 240 woods referred to above, 22 (88%) were passerines and just 3 (12%) were non-passerines. The respective figures for species recorded in less than 75% of the woods were 29 (55%) and 24 (45%). Therefore, a high proportion of non-passerines is confined to a small proportion of British woods. This may be because non-passerines include more species with exacting habitat requirements, or possibly with restricted ranges. Alternatively, non-passerines tend to be larger birds with larger territories or home ranges which may lead to wide spacing between individuals and consequent absence from some smaller woods. There is also a weak tendency for migrants to be less widespread than resident species. Of the species recorded in more than three-quarters of the woods, 17% were migrants, while among the less widespread species 28% were migrants.

The information in Fig. 3.2 tells us nothing of the ways that birds actually use woods. Some species nest and feed within woodland (for example, most of the smaller passerines) while others may nest there but feed on the surrounding land, for instance Kestrel, Stock Dove, Jackdaw and Starling. Other species may obtain part of their food from outside the wood, the Woodpigeon being a good example.

Table 3.1. *The major nest sites, feeding sites and food of woodland and scrub species. Feeding information relates to the breeding season.*

	Numbers of species[a]	
	Woodland birds	Young-growth species
Nest site		
Ground	7 (16%)	11 (79%)
Field layer	4 (9%)	3 (21%)
Shrub layer	12 (28%)	2 (14%)
Canopy	9 (21%)	0
Hole/crevice/cavity	15 (35%)	0
Feeding site		
Ground	11 (26%)	9 (64%)
Field layer	2 (5%)	4 (29%)
Shrub layer	14 (33%)	2 (14%)
Canopy	12 (28%)	0
Dead wood/bark	4 (9%)	0
Air	1 (2%)	1 (7%)
Unclassified	8 (19%)	2 (14%)
Food (adults)		
Vertebrates	2 (5%)	2 (14%)
Invertebrates	30 (70%)	8 (57%)
Plant material	7 (16%)	2 (14%)
Mixed (invert. & plant)	4 (9%)	2 (14%)
Food (chicks)		
Vertebrates	2 (5%)	2 (14%)
Invertebrates	30 (70%)	10 (71%)
Plant material	6 (14%)	1 (7%)
Mixed (invert. & plant)	5 (12%)	1 (7%)

[a]Species have been assigned to their **major** nest sites, summer feeding sites and summer foods based on information in Appendix 1 ('woodland species' are listed in part A; 'young-growth' are in part B). Species defined as 'marginal' in Appendix 1 have been omitted, giving totals of 43 and 14 species respectively. Totals under each heading may exceed these numbers because some species have been assigned to two categories; where more than two categories apply, the species has been recorded as unclassified.

Various aspects of the niches occupied by woodland birds during the breeding season are summarised in Table 3.1; the details for individual species are given in Appendix 1. The majority of the bird species that nest and feed in closed-canopy woodland and scrub are predominantly insectivorous in the breeding season. This includes all of the migrants, though some do take berries as they become available later in the summer. The main herbivorous groups are finches, buntings, gamebirds and pigeons, though many of these species raise their young on a diet of invertebrates, which may provide an essential source of protein for growth and development. Among the woodland and scrub passerines, only Greenfinch, Siskin, Linnet, Redpoll and Crossbill feed their young exclusively on seeds. In terms of numbers of species, the major feeding sites used by woodland birds are approximately equally divided between the ground, the shrub layer and the canopy, with a smaller number of species gleaning insects from dead wood or bark.

Fig. 3.3. More than one-third of British woodland bird species are hole-nesters. The Willow Tit is unique among small songbirds because it excavates its own nest-hole. This nest-hole is in an elder which is an especially important tree for hole-nesting birds in scrub habitats. Eric and David Hosking.

The two most frequently used nest sites are the shrub layer and holes. More than a third of all woodland bird species are hole-nesters. Nesting in tree-holes appears to be a good strategy for species that nest early (Nilsson, 1986a). Holes offer nest sites that are relatively safe from predators in the early part of the nesting season, presumably because there is rather little vegetation at this time of year in which to conceal open nests. Later in the breeding season, however, nests in holes may be at a disadvantage because they may be more vulner-

able to outbreaks of nest parasites. There can be strong competition for good quality holes, both among individuals of the same species and between different species. Working in southern Sweden, Nilsson (1984) found that Starling, Nuthatch and Blue Tit preferred to nest high above the ground, almost certainly because the highest tree-holes were less accessible to predators. Within each of these three species there appeared to be competition for high tree-holes because a larger proportion of the individuals nested in relatively low holes in those years when population density was high. Marsh Tits consistently nested low in Nilsson's study sites and he suggested that this was a consequence of competition with the other hole-nesting species to which the Marsh Tit was subordinate. The importance of competition in influencing the use of tree-holes probably depends on the availability of holes. Wesołowski (1989) has argued that in primeval forests with an abundance of tree-holes, competition for holes may be much less pronounced than in managed forests.

By far the greatest part of the woodland foliage occurs in the canopy, yet a surprisingly small number of species nest within it; Table 3.1 shows that just nine species (21%) do so. This is in striking contrast to the North American forests where many birds, but notably the wood-warblers (Parulidae), nest high above the ground in the canopy. For instance, half of the 22 species that are common nesters in deciduous forest in New Hampshire nest in foliage that is on average more than 5 m above ground (Holmes, 1990). The northern species of parulids are mostly small migrants. Only four of the potentially canopy-nesting species in British woods are of comparable size to these parulids: namely, Goldcrest, Firecrest, Long-tailed Tit and Siskin. In Europe, most long-distance migrants breed in the early and middle stages of forest growth (exceptions in Britain are Pied Flycatcher, Redstart and Wood Warbler) and none nests in the canopy (Helle & Fuller, 1988). This difference in habitat selection among migrants is probably related to the fact that the migrants on the two continents use rather different suites of habitats in winter. A relatively high proportion of those breeding in North America tend to winter in tropical forests. Those from Europe make far more use of savannas and other open habitats (Mönkkönen & Helle, 1989). I shall return to the subject of habitat selection in migrant birds in later chapters.

The breeding season

The months of April, May and June are those when woodland bird song and nesting are at their peak in Britain. This contrasts with tropical forests where breeding occurs throughout the year, although most species appear to have a well-defined breeding season. Food availability for the chicks is likely to be the main factor that determines the time at which birds nest in temperate woodland. For most species this may relate to the growth requirements of the young in the nest, for others it could be that an abundance of food for the recently fledged young is more important. Another constraint on the timing of breeding is that the female may be unable to find sufficient food to lay a clutch early in the season. A third factor is that nests placed on the ground or in foliage may be too exposed to predators or bad weather to allow successful nesting until the vegetation has adequately developed. That most species nest in the spring is hardly surprising given the huge flush of foliage and insect food, but there are some striking differences between species in the timing and duration of nesting seasons. The most obvious is that migrants such as warblers tend to nest later than residents. In southern and central Britain, the second half of April is the period when most resident birds lay their first eggs though some, such as Long-tailed Tit, start as early as late March. Most summer visitors lay their eggs in the first half of May. Of course, breeding activity starts well ahead of egg laying.

Many of the residents, especially tits, thrushes, Treecreeper and Nuthatch, are vigorously defending their territories during March. Their song can be intense until the end of April, but starts to decline once egg laying is complete. The rapid fall off in song is especially noticeable among those species that are mainly single-brooded, the tits and Nuthatch. The woodpeckers too become relatively silent at about the same time. As the song of resident species abates, migrants start arriving and they sing most strongly during the first half of May. By June the overall levels of woodland bird song have generally subsided substantially – many of the individuals still singing will be those that have not managed to attract a mate or that have a late nest. This pattern of events occurs later the further north one travels, and may perhaps be delayed by as much as 2 weeks in woods in northern Scotland compared with southern England.

To work out the optimum breeding season of any species requires careful study. Though nests of a species may be found over a period of several weeks, breeding performance may be closely related to the exact time of laying. In the Great Tit, early clutches (those laid in mid- to late-April) are larger than late ones, and the post-fledging survival of young from the early clutches is highest (Perrins, 1979). The fact that many individuals do not nest at the most productive time suggests that either some females may be unable to form eggs early in the season because there is insufficient food, or perhaps that attempts to nest later are made by inexperienced or poor quality birds.

In the Blackbird, however, the optimum breeding time appears to be in mid-season (mid-May) when clutches are largest and the young heaviest (Snow, 1988). Weight of the nestlings, at least in the Blackbird, is particularly significant because heavy nestlings are more likely to survive to breed than are light nestlings (Magrath, 1991). Differences between species such as these probably reflect adjustments to the availability of preferred foods. In the case of the Great Tit, the important food resource is leaf-eating caterpillars which become abundant in mid-May. The Sparrowhawk provides a clear example of how breeding seasons are linked to food. In southern Scotland, Newton (1986*a*) found that the first hawks started laying 5–10 days after the fledglings of their prey species started to feature in their diet. The period when the majority of hawks had dependent young (mid-June to late-July) matched the time when prey-fledglings were at their most abundant. It seems that the hawks depended on the surge of these fledglings to provide the extra food supply necessary for egg formation.

The length, as well as the timing, of the breeding season can be broadly related to food supply. Single-brooded species, such as the tits, have in general evolved to take advantage of one particular food resource that may be superabundant for a relatively short period. Multiple-brooded species on the other hand are more likely to exploit a range of food resources. Robin, Song Thrush and Blackbird, for example, are omnivorous and each can have more than two broods in a season. Bullfinch, Linnet and Greenfinch have equally long breeding seasons, and each is double-brooded. These finches feed their young on seeds and exploit different plants as they come into seed.

Woodpigeons have the longest breeding season of all woodland birds. The Woodpigeon breeds from early April to October, though its peak nesting period is from June to August, well after most other woodland birds have finished. This protracted season is possible because the Woodpigeon is a supremely versatile feeder, taking a succession of leaves, flowers and fruits as they become available. This vegetable matter is even supplemented by some invertebrates, including caterpillars and small molluscs.

For species with long breeding seasons few, if any, individuals breed throughout the entire potential season. Exactly why some individuals breed early and some late is something of a mystery – in some cases the older or more experienced birds may choose the 'best' time. What is clear is that in any wood only a fraction of the entire breeding population of birds will actually be nesting at any particular time. As the breeding season progresses, there is substantial flux both of species and individuals. The notion of a fixed number of breeding pairs in a particular wood may be unrealistic for many species. Some birds may cease breeding early in the season while others may establish late territories, but this can be detected only by catching and individually marking the entire population of a species within a study area. When this was carried out for Willow Warblers living in an area of scrub in Surrey, a fairly high turnover of male birds was discovered (Lawn, 1982). Out of the total 172 territories located over 6 years, 76% were occupied for the entire breeding season, 4% were second territories established by males, 8% were temporarily held by males which disappeared before the main breeding season got underway, and 12% were set up late in the season. Successful breeding took place in few of the temporary or late territories.

A further complication in gaining a realistic picture of 'the breeding bird community' of a wood is that by no means all birds breed in monogamous pairs, as assumed by some census techniques. In fact,

Fig. 3.4. The Robin is one of eight species that are virtually ubiquitous in British woodland. In small woods, there may be substantial turnover of breeding individuals even within a single breeding season (S. Hinsley personal communication). Chris Rose.

the number of species that exclusively nest in pairs may be rather small. The more research that is carried out, the more it seems that many species have a proportion of individuals that indulge in polygamy of some form. In the above study of Willow Warblers, 11 cases of bigamy were recorded, though previously this behaviour was thought to be extremely rare in the species. Among other woodland species proven to be frequently polygamous are Woodcock, Dunnock, Pied Flycatcher and Wood Warbler. The latter is often polyterritorial, whereby a single male simultaneously defends more than one discrete territory (Temrin et al., 1984). These comments are made here partly to emphasise the complexity of determining exactly what constitutes the breeding bird community of a wood. Most censuses and counts of breeding birds, even when extremely well designed and executed, can give only approximations of true numbers of birds, or indices which can be used to compare the relative numbers from one place or year to another.

Periodically, many woods experience massive increases in numbers of foliage-feeding caterpillars leading to defoliation of the canopy. In Britain this phenomenon is most obvious in oakwoods. Species such as Jay, Jackdaw and Woodpigeon, as well as the smaller insectivorous species, all exploit this superabundant food supply. The effects of such insect outbreaks on bird populations are complex (L. Tomiałojć personal communication). Breeding success of some species, like Chaffinch, can be much higher in defoliation years than when there are relatively few caterpillars. In part, this success is a direct response to a better food supply. But nest losses may be low in defoliation years because predators, such as woodpeckers and Jays, may also be feeding heavily on the caterpillars with the result that they take fewer eggs and nestlings of songbirds. Conversely, some canopy-nesting birds, for example Hawfinch, may suffer increased nest losses in defoliation years because their nests may become more exposed to predators. It is unlikely that birds exert any controlling influence on such large-scale outbreaks of insects but, at lower densities of prey, it does appear that birds can reduce the numbers of woodland insects in the breeding season (Holmes et al., 1979; Nilsson et al., 1985).

Post-fledging and autumn

A walk through the woods in late July or August may lead one to think that they are almost devoid of birds. But this can be an illusion, for birds can be very inconspicuous at this time. Many are moulting and are potentially vulnerable to predators, especially Sparrowhawks. Exactly how many of the breeding birds remain in the woods at this time is extremely difficult to ascertain because they are so hard to observe, but it is likely that a high proportion of the resident species do so.

The weeks following their departure from the nest are crucial ones for young birds. The risks from predation for inexperienced juveniles are high. The young also have to make the difficult transition to independence which requires gaining competence in feeding, finding suitable roost sites and so on. Most of the losses of young birds probably occur very soon after they have left the nest, but the survival rates of most young small passerines probably remain lower than those of adults until at least late autumn.

There are other reasons why the period following breeding may be especially important in the life cycle of some birds. This is the time for young birds to become familiar with their habitats: certain locations and vegetation types will become imprinted on them and this will influence their choice of breeding site when they come to breed. Most woodland passerines appear to show relatively low breeding dispersal – the movements made by adults between successive breeding sites. Adult Great Tits, for example, return more or less to the same territory. Natal dispersal, which is the movement of young between birth place and their first breeding site, is more variable but typically is greater than breeding dispersal. In some species, juveniles breed quite close to their birth place. One study of Great Tits suggested that the young settled only some four to seven territory widths from their birth place, although the number of birds that left the wood was not estimated (Greenwood *et al.*, 1979). The evidence would suggest that some juvenile warblers, for instance Willow Warblers (Lawn, 1982), show very low rates of return to the site where they were reared. Species that live in rapidly changing habitats, such as young coppice, plantations and scrub, may have a greater need to explore for alternative breeding sites. This would include many of the migrant species. One might expect that these birds should show higher levels of natal dispersal than those which live in more stable, older woodland habitats. This idea has yet to be tested properly. Nonetheless, before migrating south, some species of migrants are known to undertake fairly extensive movements in the post-fledging period. These could conceivably be related to birds searching for new potential breeding sites or creating 'navigational targets' which enable them to relocate these sites (Baker, 1993). Juvenile Willow Warblers, for instance, disperse in all directions, perhaps up to 25 km from the breeding site in the latter half of their moult period (Lawn, 1984).

A conspicuous feature of the immediate post-fledging period is the rapid formation of mixed flocks of *Parus* tits, in which several family

Fig. 3.5. A mixed flock of tits. Flocking in autumn and winter is probably advantageous to these birds in helping to reduce the risks from predation and it may also help the birds to forage more efficiently. Chris Rose.

parties of various species may come together, often joined by Long-tailed Tits, Goldcrests, and Treecreepers. The flocks can be found at all times outside the breeding season. There are two main benefits from feeding in flocks. One is that birds in flocks may find food more easily; they may achieve this by copying the successful individuals, either by adopting particular feeding techniques or by switching to a location that appears to hold more food. Such advantages may be especially strong if the food resource is patchily distributed. The second is that staying in flocks helps to reduce the risk of being caught by a predator. Many pairs of eyes are likely to detect a Sparrowhawk far sooner than a single pair so that an early warning can be given. Also, the chance of an individual being singled out by a predator decreases as flock size increases. A reduction in predation risk may allow birds to devote less time to vigilance and more to feeding. It has been suggested that the Treecreeper benefits in the latter way from associating with tit flocks (Henderson, 1989).

Autumn brings a resurgence of territorial activity in several woodland species which essentially live on territories all year round. Among these species are Robin, Nuthatch and Great Tit. As explained earlier for the Nuthatch, such autumn territoriality may regulate the numbers of birds that breed the following spring, though this seems uncertain in the case of tits (Perrins, 1979).

In late summer and autumn, seeds and fruits become more widely available. At this time many species, including those which are predominantly insectivorous during the breeding season, cash in on this easily gathered food, both in woodland and scrub (Snow & Snow, 1988). Some sites, especially in scrub habitats (Chapter 5), can become 'feeding hotspots' attracting many birds which have not necessarily bred in the surrounding area. A few species routinely store seeds at this time of year – two of the most studied are Marsh Tit and Jay. The latter is especially interesting because it has evolved a symbiotic relationship with the oak (Bossema, 1979). Individual Jays will store literally thousands of acorns, mainly during October. This enables them to exploit acorns as a staple food in most months of the year. Many of the acorns that are not recovered may germinate and thus the Jay plays a major part in the survival and dispersal of the oak. Through eating their fruits and berries, birds act as dispersal agents for many species of trees and shrubs, but in few cases has such a close interdependence evolved as between the Jay and the oak.

Winter

In most British woods, the numbers of birds are probably at their lowest in mid- or late-winter. This is certainly the case in woods in northern and central Europe where the climate is too severe for many species to overwinter. Familiar winter birds in Britain, such as Wren, Dunnock, Robin, Blackbird and Chaffinch are summer visitors across a large part of their European range. Many upland woods and those in northern Britain lose some species that are otherwise more or less resident in other parts of the country. Robin and Chaffinch, for instance, are absent or much reduced in numbers in upland woods in winter, and may even abandon some lowland woods (Beven 1976).

The seasonal pattern of change in many British lowland woodland bird communities is probably similar to that described for oak shelterwood forest in Burgundy (Frochot, 1971). Here, bird communities in the oldest stands of 150 to 200 years, were relatively stable compared with those in young-growth of less than 20 years. The young-growth showed a strong peak in overall bird density in the breeding season, largely due to the strong selection of these stands by migrant warblers. A smaller peak occurred in autumn, mainly due to influxes of finches presumably attracted to seeds. The smallest bird populations were present between December and February. In the old stands there was no marked peak of bird numbers in the spring, neither was there such a striking trough in winter, though numbers were lowest in January. Overall bird numbers in the two stages of tree growth were similar in the breeding season, but in winter the old shelterwood supported far more birds than the young stands, though the counts did not appear to take account of roosting birds.

Many resident woodland birds in Britain remain in the vicinity of their breeding areas throughout the winter. Nonetheless, there is a tendency for them to become more widely distributed and to use a greater range of habitats outside the breeding season (Bilcke, 1984; Lack, 1986). Only in a few species – Nuthatch, Willow Tit and Marsh Tit – do most individuals remain territorial on their breeding sites throughout the winter. Great Tits spend at least the early part of the winter in territories, though they will readily join tit flocks as the winter progresses and leave their territories in hard weather. Other tits tend to winter mainly in flocks, which may roam outside the woodland especially in hard weather. Among some species, notably Robin, Blackbird, Wren, Treecreeper and Goldcrest, only a proportion of the birds appears to hold territory through the winter. In winter, Robin and Wren will use habitats, such as reedbeds, that are quite unsuitable for breeding. In the British Robin, the sexes hold separate winter territories with the males remaining on the breeding site. In Belgium, it seems that many Robins vacate woodland in winter, although they do maintain winter territories in gardens and parkland (Adriaensen & Dhondt, 1990*b*). Some species are not at all territorial in winter. A good example is Chaffinch, which largely abandons the woods for feeding, except when there is a good crop of beechmast. The birds typically feed in small flocks in the vicinity of the breeding site.

Winter food and foraging sites of many woodland birds can be substantially different from those used in the breeding season. Many resident species that are essentially insectivorous in summer will eat large numbers of seeds in winter. The finches are all granivorous in winter while Dunnock, Robin, tits and Nuthatch eat both insects and seeds. Winter is the main time that Great Spotted Woodpeckers excavate dead wood searching for saproxylic insects (those living in dead wood), while in northern woods they appear to rely heavily on pine seed. In summer, however, the woodpeckers feed heavily on defoliating insects. As the winter progresses, some species can undergo quite subtle changes in feeding behaviour. A good example is offered by Great Spotted and Three-toed Woodpeckers feeding on dead or decaying spruce trees in southern Norway (Hogstad, 1971). Great Spotted Woodpeckers generally fed higher than Three-toed Woodpeckers but the greatest overlap in feeding heights occurred in October and March, with the least overlap in mid-winter. Weather can affect the sites used by birds feeding in woodland. On cold days, some species tend to feed at lower heights than on warm days, while other species may do the opposite (Bilcke *et al.*, 1986). Such changes in feeding sites within the winter may well be driven by competition between species.

Woodland and scrub provide roost sites for species that feed mainly in other habitats during winter. These species include Starling, thrushes and finches. Sheltered roost sites are important for at least two reasons: to provide cover from predators and to help in the reduction of overnight weight loss, which occurs as a consequence of the need to maintain body temperature. Many species roost communally; perhaps the most spectacular are Bramblings, which on the Continent can number millions in areas with heavy beechmast crops. These birds choose roost sites, often young conifer woods, which are well sheltered from the wind (Jenni, 1986). Scrub and young conifer plantations are particularly favoured roost sites for many finches, buntings and thrushes. Rhododendron thickets are much used by finches. Presumably these habitats are selected because the dense vegetation offers good cover from both weather and predators. Communal roosting may come about partly because good roost sites

Fig. 3.6. Starling roost. Though the numbers of birds feeding in woodland and scrub in winter are generally lower than at any other time of year, these habitats support huge numbers of roosting Starlings, thrushes, finches and buntings. Especially important roost habitats include dense scrub, thicket-stage conifer plantations and rhododendron. Chris Rose.

are scarce, causing birds to congregate in a small number of places. This seems inadequate as a general explanation and birds may gain two benefits from roosting in large numbers. First, as with feeding in flocks, the chances for the individual bird of being preyed upon are probably reduced. Secondly, roosts could act as 'information centres' whereby birds learn the location of productive feeding sites by following individuals that fed successfully the previous day. This supposes that birds can identify the successful individuals and that, over a period of time, such benefits will accrue to all individuals. This attractive idea would be most likely to apply to species that dispersed widely and fed on patchily distributed food. Unfortunately though, there is little firm evidence to suggest that roosts do serve this function (Mock *et al.*, 1988).

4

Abundance and distribution of woodland birds

Some patterns in the distribution and abundance of woodland birds are obvious. An example is the contrast between bird communities of Welsh upland oakwoods containing Pied Flycatchers, Redstarts and Wood Warblers, and those of adjacent mature spruce plantations that lack these birds but carry many more Goldcrests and Coal Tits. On the other hand, many spatial patterns are far less easily detected but nonetheless real. Some species show non-random distributions within apparently fairly uniform woodland that are probably responses to subtle variations in habitat quality that are difficult to measure.

There is no such thing as a 'typical woodland bird community'. Enormous variation exists in the abundance of individual species, the total density of birds and the numbers of species found in woods. Published bird census studies indicate that the densities of birds breeding in British woodland commonly range from 200 to 1600 territories per km^2 (Petty & Avery, 1990).

This chapter is about principles; it reviews the main determinants of woodland bird communities. In doing so, it is important to recognise that the numbers, types and abundances of species present in any wood, or part of a wood, will be the product of many factors. It would be misleading, for instance, to attempt to explain woodland bird communities solely in terms of vegetation structure or of tree species. Inevitably there is interaction between the different characteristics of woods and it can be immensely difficult to disentangle the effects of one factor from another. To take one example, woodland structure is itself affected by the dominant tree species. Furthermore, different factors operate on different scales; some apply at the level of the entire wood (e.g. geographical location, area), others can be far more local, often varying appreciably within the wood (e.g. structure, tree species).

The factors shaping woodland bird communities are considered individually in this chapter; they fall into five broad groups. Geographical location, altitude and land productivity are essentially *invariable* for they are less readily altered by human activity than the other factors, though this is not strictly true of land productivity. Woodland area and isolation can be thought of as *landscape* attributes. Edge effects, growth stage, patchiness and stand structure are aspects of the *physical structure* of the woodland vegetation. Tree species composition is concerned with the *floristics* of the vegetation. Finally, *behavioural and demographic* attributes of the bird species themselves, such as social attraction, population size and dispersal ability, may be important forces determining local distributions. To this last group one could add interactions between the bird species, especially competition, though these are not dealt with here because they were considered in the previous chapter.

Geographical location

Woods in northern and western Britain tend to be the poorest in species (Fuller, 1982). On average, woods in north-west Scotland hold the very lowest numbers of species. Woods in west Wales are poorer in species than woods at a similar latitude to the east. These declines in numbers of species towards the north and west occur in most groups of birds; they do not appear to be a consequence of certain groups of birds being strikingly under-represented in these regions (Fuller, 1982). This pattern appears to be part of a wider geographical effect. The number of breeding bird species in Ireland is lower than that in Britain, while several woodland species absent from Britain are typical of the Continental land mass to the east (Chapter 2).

The decline in numbers of species towards the north and west is underpinned by two facts. First, several species are confined to south and/or east Britain. These include Turtle Dove, Lesser Spotted Woodpecker, Nuthatch and Nightingale. Secondly, many of the widespread species are more patchily distributed in Scotland, occurring in a lower proportion of Scottish woods than is the case further south – for instance Chiffchaff, Garden Warbler and Blackcap. By no means all woodland species with restricted ranges are, however, confined to the south and east. Buzzard, Pied Flycatcher and Wood Warbler are strongly characteristic of western oakwoods but not of the Midlands or East Anglia. Capercaillie, Crested Tit and Scottish Crossbill are local to parts of the Scottish Highlands and Red Kite is local to Wales. Thus, location within Britain is a primary determinant of both the richness and the type of bird community that can be found in a wood.

Why should these regional trends exist? There are at least three possible explanations, some of which may be related to the large climatic differences that exist within Britain. The distribution of soil types is one possibility. The higher altitudes in Britain are mainly in the north and west, and these regions are generally associated with the less productive soils which are known to support lower populations of many birds (see below). If soil type influences total bird density, it probably influences species number in a similar manner because over-all density and number of species are usually closely related. Secondly, insect food for many birds may be less abundant in the relatively cool and wet climate of the north and west. Indeed, a combination of land productivity and climate may act to reduce the primary pro-ductivity of woods in the west and north. In a study of birchwoods in the Scottish Highlands, Bibby *et al.* (1989*a*) found that the density of birds and numbers of species declined from east to west. They interpreted this gradient in terms of diminishing productivity towards the west, where the production of foliage and abundance of insects may have been lower than in the east. A third factor is that northern woods, especially Scottish woods, are more frequently dominated by conifers than woods elsewhere. In general, coniferous woods support lower numbers of species, and lower densities of birds, than broad-leaved and mixed woods (see below).

Altitude

Woods at higher altitude are generally poorer in bird species and overall numbers of birds than lower-lying woods. Decreases in num-bers of breeding bird species associated with increase in altitude have been demonstrated in France, Germany and North America (Able & Noon, 1976; Lebreton & Broyer, 1981; Wink & Wink, 1986). In Britain and in Germany, within similar types of woodland, the densities of woodland songbirds are lower at higher altitudes (Newton *et al.*, 1986; Wink & Wink, 1986). The explanation of these altitudinal trends undoubtedly lies with an inter-related set of factors. With increase in altitude, the climate typically becomes harsher, the soils less pro-ductive, the number of tree and shrub species fewer, and the vege-tation less luxuriant. Almost certainly, the amount of insect food will be less, and birds may also forage less efficiently, because of the poorer weather. It is likely that the breeding productivity of many woodland birds declines with increasing altitude. This is the case for the Nuthatch in the Harz Mountains where the species breeds up to a limit of about 650 m, some 400 m below the treeline (Zang, 1988). A decline in Nuthatch density with elevation is matched by a trend of later breeding, smaller clutch sizes and lower breeding success.

Land productivity

When all other factors are equal, the density of birds in woodland increases with the fertility of the soil. In Finland, within the same woodland types (birch, spruce or pine), the density of birds is consistently greatest on the most productive soils (von Haartman, 1971). Results collected by Newton *et al.* (1986) as part of a study of the food available to British Sparrowhawks, suggested that the abundance of songbirds was closely linked with soil productivity. Presumably these trends are driven by larger amounts of insect and plant food being available for songbirds in the richest habitats.

Woodland area

A widespread pattern in nature is that more species live in larger patches of habitat. This applies to many groups of organisms, including birds in woodland. Many studies, both in North America and Europe, show that large woods support more species than small woods. Strong relationships between the size of British woods and the numbers of bird species they support have been demonstrated by Moore & Hooper (1975), Woolhouse (1983), Ford (1987), Fuller (1987) and Hinsley *et al.* (1992). Examples are shown in Fig. 4.1 for three

Fig. 4.1. Three British examples of relationships between the number of breeding bird species supported by a wood and the area of the wood. Open circles represent more than one datum point. The three studies differ in the sizes of woods examined: (a) Ford (1987), (b) Woolhouse (1983) (with permission from Elsevier Science Ltd), and (c) Fuller (1987).

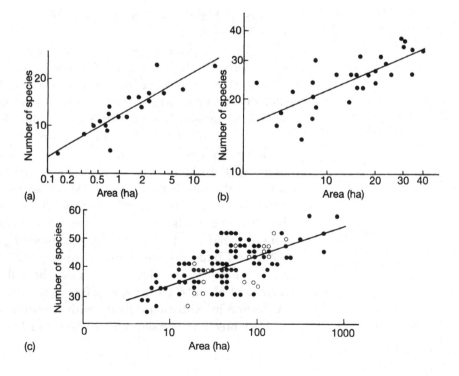

sets of woods, each covering a different range of sizes. The exact rate at which species are added as the area increases may depend on the sizes of the woods studied, the thoroughness with which they were sampled, and the region. The bird species composition can also vary with woodland area (Fuller, 1982; Nilsson, 1986*b*). In Britain, it seems that several species tend to avoid the smallest woods, for example Nightingale, Chiffchaff, Marsh Tit and Jay (S. Hinsley, personal communication). Conversely, no species are confined to small woods. The species–area relationship in British woods is essentially a result of more species being added to a core of ubiquitous species as the area increases.

In addition to the number of species, two other attributes of bird communities are affected by woodland area; these are the stability of bird communities and the density of birds. Working in small farm woods mainly in the East Anglian fens, Hinsley *et al.* (1992) recorded that very small woods, of less than about 2 ha, underwent much larger year to year changes in bird species composition than did larger woods. Moreover, the turnover between years in individual adult birds was very high in these small woods in species such as Wren, Robin, Dunnock and Chaffinch. This suggests that bird populations in these woods were being sustained by immigration. In contrast to the number of species, the overall density of birds appears to be lower in large than in small woods (e.g. Nilsson, 1986*b*; Ford, 1987). This effect probably arises because small woodland patches have relatively large amounts of edge habitat which often carry higher densities of birds than interior woodland (see below).

The description of species–area relationships tells us nothing about the processes underlying them, although this has been the subject of strong debate. There are five main candidates. MacArthur & Wilson (1967) proposed that the number of species on an island resulted from a dynamic equilibrium in which rates of immigration balanced those of species extinctions. Extinctions are considered to decrease with increasing area, and immigrations to decrease with greater isolation. This equilibrium theory has been widely assumed to apply to both real islands and habitat islands, such as patches of woodland in open country. Several attempts have been made to apply it to the design of nature reserves. Implicit in the theory is the assumption that area influences the number of species through the size of the population supported and hence the probability of extinction. In other words, by chance alone, populations are expected to become extinct more often in small woods.

A second explanation arises from the idea that some species may have a minimum area within which all their needs can be met. This effect would contribute to the relationship between species number and area because a proportion of species would be excluded from the smallest woods. There is a broad relationship between the body sizes of birds and the sizes of their territories or home ranges, so in general one would expect the larger species to occupy large woods more frequently than small woods. However, it is not clear that this is the case (see Fig. 5.3 in Fuller, 1982). In any case, it is difficult to equate territory size with the notion of minimum area requirement, because the territory sizes of many species are somewhat variable, depending on habitat quality and population density.

Species that avoid woodland edges are more likely to be accommodated within larger woods, thus theoretically contributing to higher species richness in large than small woods. In the woodlands of eastern North America, there are a considerable number of forest interior species; Askins *et al.* (1987) listed 20 for Connecticut forests. This contrasts with the situation in European woodland where it is hard to identify a single genuine forest interior bird, with the possible exception of Capercaillie. Small woods may be unsuitable habitats for some species because nest predation can be much higher at the extreme edge of the forest than in the interior. This is the case for ground-nesting birds in northern forests (Andrén & Angelstam, 1988). I return to edge effects below.

The fourth possibility is that as the area increases so does the number of habitats and hence the number of species. A large area is likely to contain suitable habitats for more species than a small area. This effect can be seen in a simple way in many plantations. Within a plantation of some 100 ha, a randomly chosen block of, say, 3 ha is likely to be composed of just one or two ages of trees, whereas the entire plantation may embrace many different growth stages providing habitats for a far wider range of species. I have pointed out previously (Fuller, 1982) that several species typical of scrub or young-growth woodland occur more frequently in large woods. This may be because larger woods are more likely to have glades, rides and shrubby areas.

The final explanation is increasingly accepted as a general explanation of species–area relationships. The larger the wood, the more individual birds it will tend to contain, though, as explained above, the density of birds is usually highest in small woods as a result of edge effects. A large sample of individual birds is more likely to contain a large number of species than a small sample, purely by chance. Therefore, large woods can be thought of as offering relatively large samples of the species pool. It is hard to dispute that this effect, dubbed *passive sampling* (Connor & McCoy, 1979), is likely to contribute to most species–area relationships.

A huge effort has been expended trying to understand how species–

area relationships may help to design so-called 'optimal nature reserves', or conversely to predict the effects of forest fragmentation. There are considerable limitations with this approach because it takes little account of the real habitat needs of species and the dynamics of populations. As this chapter shows, birds respond strongly to many attributes of woods other than area. Indeed, where birds are related to area it may be indirectly through factors such as edge effects. An illustration of the complexity of this issue is given by Lynch & Whigham (1984) who found that densities of many bird species in Maryland forests were more closely related to characteristics of the vegetation than to forest area, and that species differed greatly in the factors predicting their distribution.

Woodland isolation

There are several aspects of isolation: the distance to the nearest wood, distance to the nearest extensive forest tract, the number of linking features (e.g. hedges) between woods and the overall amount of woodland in the region. Theoretically, the occurrence of some species in woods could be linked to isolation; the more isolated a wood the less likely is a species to colonise it. This is particularly true for species that do not disperse far, such as Marsh Tit. Hence, isolation could be an especially relevant influence on bird communities in small woods where rates of species extinction may be highest (see above). Bird communities are less likely to be affected by the isolation of woods than are those of less mobile groups of plants and animals. For example, few species of birds seem totally deterred from nesting in the scattered small woods that are found on the Fenland inland from the Wash, despite these being among the most isolated woods in Britain. These woods do, however, lack a few species, such as Nuthatch and Marsh Tit and it may be no coincidence that these are rather sedentary species.

There is some evidence from The Netherlands that bird communities in small woods do vary in complex and subtle ways according to their isolation (van Dorp & Opdam, 1987). In this case, the number of breeding species associated with mature broadleaved woodland appeared to be linked with isolation, but the total number of breeding species was not. In The Netherlands there are 15 such species including Great Spotted Woodpecker, Lesser Spotted Woodpecker, Redstart, Treecreeper, Nuthatch, Marsh Tit and Hawfinch that depend on mature woodland (Opdam *et al.*, 1985). Isolation has also been reported to influence bird communities in North American forests (Lynch & Whigham, 1984; Askins *et al.*, 1987).

In general, as far as British woodland birds are concerned, effects of isolation are secondary to those of woodland area and habitat variation, and can be convincingly demonstrated only when these other overriding factors are controlled or removed. Nonetheless, the potential significance of isolation for birds like Nuthatch and Marsh Tit should not be dismissed.

Edge effects

An edge effect occurs when the numbers of a species increase or decrease at the interface between two different types of vegetation or habitat, one of which may be open space. There are three types of edges that can potentially influence the distribution of birds within a wood: the edges of rides or glades, the boundaries between forest compartments (i.e. areas at different stages of growth), and the exter-

Fig. 4.2. The Nuthatch is one of the most arboreal of British birds. In some regions it is absent from apparently suitable habitat and it is possible that this rather sedentary species does not readily colonise isolated woods. Chris Rose.

nal edge of the wood. Each of these is discussed separately below; for more detail see Fuller (1991), Fuller & Warren (1991).

Widening woodland rides and creating linear strips of grass, herbs and shrubby vegetation along them is conventional conservation practice in lowland woods (Warren & Fuller, 1993). This can benefit many invertebrates and plants but the responses of birds are less clear. In summer, the species most consistently associated with the edges of rides and glades in lowland broadleaved woods is the Chiffchaff. Tree Pipits, Willow Warblers and Nightingales can also be linked with ride systems though, in many woods, rides and glades appear to have little, if any, influence on bird distribution. In general, those rides with wide marginal strips of scrub or shrubby vegetation are likely to be the most attractive to birds. Rides with sharply defined margins, where the trees directly abut the track itself, are the least attractive.

There is evidence from Sweden that edge effects exist between forest compartments of different ages. Numbers of woodland birds such as Chaffinch, Robin and Goldcrest are higher in mature conifer forest adjacent to a clear-fell than in the interior of the forest (Hansson, 1983). Tree Pipits on the other hand avoid the mature forest but are commonest at the edges of clear-fells, while Whinchats are most numerous in the centre of clear-fells. There are two likely explanations for these effects. One is that the shrub layer may be more strongly developed at the edge of a compartment than in the interior. The other is that there could be more food in the form of insects and berries at the edge. An implication of such edge effects is that the structure of the forest, in terms of the sizes and shapes of compartments, could influence the overall distribution and abundance of many bird species.

Several breeding bird species strongly prefer certain external edges, particularly in fairly uniform mature woods (those lacking young-growth such as developing plantations and coppice). This is by far the most pronounced type of edge effect shown by birds in lowland British woods. The birds associated with edges include most of those that use young-growth within woodland, for example Garden Warbler, Blackcap, Willow Warbler, Chiffchaff and Dunnock. In addition, some residents not especially associated with young-growth habitats, commonly reach their highest densities at the external woodland edge among them Wren, Chaffinch, Blackbird and Long-tailed Tit. In two studies of bird distribution within English woods, 40% and 67% of breeding bird species were associated with the woodland edge (Fuller & Warren, 1991). An example from one wood is shown in Fig. 4.3 where the territories of four species are mapped. Willow Warbler is confined to the extreme edge of the wood. Although less pronounced, Chaffinch and Wren also show a preference for the edge, but Robin appears to be randomly distributed within the wood. There are no species in lowland English woods that appear to avoid the woodland edge, though it is possible that, historically, some species did so before clearance of the natural forest became extensive.

Why should birds prefer to nest near some external woodland edges? One of the most obvious reasons is that the vegetation structure at the edge is often different from that in the interior. The shrub layer is often denser at the edge, perhaps because of the extra light (Fuller & Whittington, 1987). It can be seen in Fig. 4.3 that not all edges are equally attractive for birds. In part, this may be a consequence of vegetation structure, for those edges with few birds often have relatively little in the way of shrub layer. It is also possible that north-facing edges are the least attractive to birds because these are relatively heavily shaded and may be poorer for insects, though there seems to be no published evidence to support this idea. The communities of shrubs and trees at the woodland edge are often different, for

Fig. 4.3. The distribution of Wren, Willow Warbler, Robin and Chaffinch territories within a derelict coppiced wood in Buckinghamshire. The data were collected by the author in 1984.

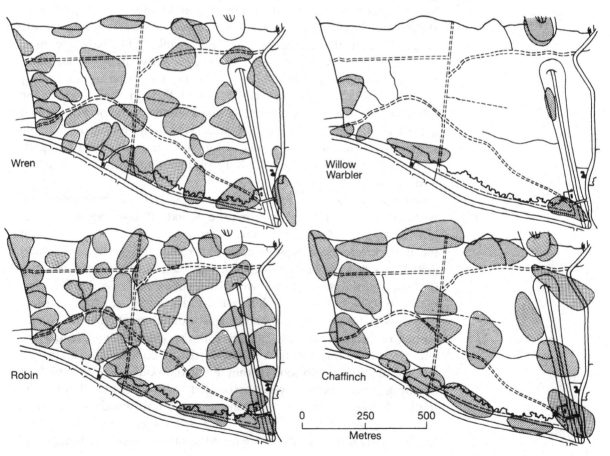

instance richer in species, from those of the interior. This presumably leads to greater amounts of insects, seeds and berries. Food resources at the edge will be further enhanced because many shrubs flower more regularly there than in the interior, and the light, warm conditions at the edge benefit many insects. External edges may offer more suitable habitats for birds than internal edges because they tend to be older. Many rides and compartment boundaries are relatively recent. Both the structure and the plant composition will be more complex along old, established edges where there has been a constancy of light penetration (Fig. 4.4).

Birds that feed in open country but nest in woodland are likely to nest near the edge. Among such birds might be Starling, Rook and several finches. This could contribute to marked differences in composition of bird communities at the edge. This is the case in Poland's Białowieża National Park (Tomiałojć & Wesołowski, 1990). Many of the species that select the edges of British woods do not feed on surrounding farmland – many feed on insects in the woodland foliage – so this is not a satisfactory general explanation of edge effects.

Growth stage and patchiness

Woodland is dynamic. Plantations and coppice go through more or less regular cycles of cutting and regrowth. In the absence of human intervention, all woods would go through longer cycles of development, maturation, death and regeneration. These natural cycles occur at scales ranging from the small gaps created by the demise of individual trees to swathes of storm damage extending over many hectares (see Chapters 2 and 9). The term *succession* is often applied to the development of plantations and coppice where one stage of forest growth gives way to another but the species composition of the trees often remains broadly constant. The ecological concept of succession is strictly concerned with more natural changes in vegetation which may involve one community of plants being gradually replaced by another. A treefall gap, for example, may initially be colonised by birch and sallow, which in time, might give way entirely to ash and subsequently oak. To what extent this difference between the development of plantations and natural forests affects birds is unknown. What is obvious, however, is that very different communities of birds occur in the various stages of both plantations and natural forests.

As a general rule, the numbers of species and overall density of birds in European forests tend to increase as the trees grow (Helle & Mönkkönen, 1990). An exception is found in coppice woodland where numbers of species and densities sometimes decrease in the oldest stages (Chapter 6). Some conifer plantations also appear to carry greater densities of breeding birds in the thicket stage than in mature forest (Constant *et al.*, 1973). Many species of birds clearly prefer a particular stage of growth. Some are confined to the earliest stages

Fig. 4.4. The edge of an oakwood in mid-Buckinghamshire. The highest densities of breeding birds in this wood are found close to this edge. The vegetation is extremely dense, partly as a result of vigorous blackthorn thickets, and there is a high diversity of shrubs and trees along this edge. Rob Fuller.

when the vegetation is open, others to the period when the canopy is starting to close and the vegetation is very bushy. Other species, including most hole-nesters in the absence of nest boxes, select those stages with mature trees. Examples are given in Chapters 5, 6 and 8 where responses of bird communities to the development of scrub, plantations and coppice are discussed. A general point to be made here is that because species differ in their preferred stages of growth, woods with the greatest variety of growth stages provide more habitats for birds, and hence support more species, than uniform woods. Additional patchiness in woodland habitats can be created by rides, glades and treefall gaps. These sometimes provide the only suitable habitat in a wood for species such as the Tree Pipit.

Fig. 4.5. The Whinchat is characteristic of upland conifer plantations, but it only occurs in the youngest stages of growth. Chris Rose.

Stand structure

Woodland structure is often thought to be a primary factor influencing woodland bird communities. This is undoubtedly true at one level. Increases in the density of birds and numbers of species as woods grow and mature, is a response to the structural development of the habitat. In broad terms, the more complex the vegetation, the more complex the bird community. Studies in North America and Europe have shown that the bird community becomes richer as more layers of foliage are added (MacArthur & MacArthur, 1961; Moss, 1978). These studies measured habitat structure in terms of an index called *Foliage Height Diversity* (FHD); high values of FHD are found in woods with many layers of foliage. The idea that the vertical foliage profile is the main factor determining the richness of birds sounds straightforward and appealing. In fact, some biologists have even suggested that such structural effects are the only important ones in explaining variations in woodland bird communities and that factors such as the tree species composition are irrelevant. However, several studies have been unable to confirm any link between birds and FHD, while others have found that different measures of habitat structure, such as crown volume, are better predictors of the number of bird species (Verner & Larson, 1989). These apparent contradictions partly arise because people have studied the relationships between bird communities and habitat structure using fundamentally different selections of study sites (Fuller, 1994). Where the comparison has been between different growth stages, or habitats varying greatly in physical appearance, it is hardly surprising that the effects of woodland structure have been overwhelming. The importance of structure has often been less striking when the study has examined woods of the same growth stage or similar habitats.

Some further discussion of the effects of the vertical and horizontal aspects of woodland structure is in order at this point. Woodland birds differ considerably in the heights at which they forage. The canopy, the shrub layer and the field layers each have their own species (Fig. 4.6). There is also much variation between birds in the heights at which they nest and the types of nest sites they prefer (Appendix 1). So it is not surprising that, all other things being equal, woods with more layers of foliage tend to be the richest in birds for they offer feeding and nesting sites for a wider range of species. Dense foliage is also important in offering shelter. Roosts of finches, thrushes and Starlings are usually located in thick shrubby vegetation. In Swedish beechwoods, Nilsson (1979a) found that when birds of prey were common, wintering tits and Nuthatches were more abundant in areas with a reasonable cover of trees and bushes below 8 m, which probably offered some protection from the predators.

The structure of the shrub and field layers is especially important to many species. In general, woods with sparsely developed shrub layers are poorer in birds than those with a dense shrub layer (Figs. 4.7 and 4.8). An exception occurs where rhododendron or holly forms the understorey; few breeding birds seem to benefit from dense growth of these shrubs, though it may form a good roost site. Some warblers, most strikingly Blackcap and Garden Warbler, prefer woods with a continuity of dense foliage from ground level to a height of at least 2–3 m. Other species, such as Wren and Chiffchaff, tend to select areas with a dense field layer. Low mats of bramble, for example, are widely used by both species for nesting. The beechwoods of the Chilterns offer a good example because the mature stands range in structure from those totally devoid of any field or shrub layer, to woods with well developed growth of bushes and small trees. The latter woods typically have moderate numbers of thrushes, warblers and Wrens, which are more or less absent from the most open beechwoods. Not all species select woods with dense undergrowth. One warbler, Wood Warbler, generally avoids woods with a dense

Fig. 4.6. Vertical distribution of selected species during the winter in Bagley Wood, Oxfordshire. Relative abundance in different height zones is indicated by the thickness of the individual diagrams. From Colquhoun & Morley (1943) with permission from Blackwell Scientific Publications.

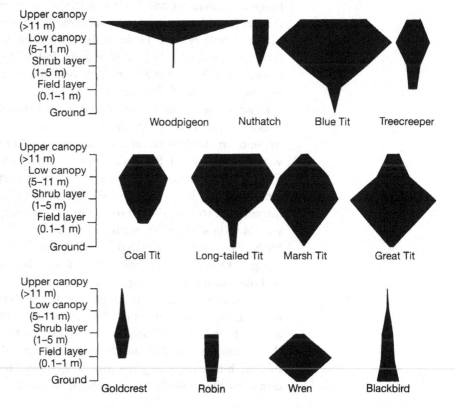

shrub layer, as do Pied Flycatcher and Tree Pipit. For birds that forage on bark or in the tree canopy, the shrub layer may be irrelevant.

A sparse shrub layer usually arises either as a result of heavy grazing and browsing (Chapter 7) or because the dominant trees cast an extremely heavy shade. Dense shading is a property of some trees more than others; beech and sycamore cast heavy shade compared with ash and birch. But canopy openness – the horizontal element of stand structure – is also affected by management. Left to its own devices, the canopy of most woods will gradually become more open as trees die and gaps form. In the absence of severe grazing, this will stimulate regeneration and the growth of shrubby vegetation. These gaps may then be colonised by warblers and other birds needing a dense shrub layer. Furthermore, insects and fruits can be more abundant in gaps so that the food resources for birds can be higher there than under the closed canopy (Blake & Hoppes, 1986). Thinning of stands can have a similar effect to natural gap formation, by releasing shrub growth, which can lead to increases in numbers of warblers (Fuller, 1990). Hence, there can be considerable interaction between the horizontal structure of a stand and its vertical structure.

Structural 'micro-features' of vital importance to some species are holes and dead wood. Their availability is strongly influenced by tree species and by management practices. Nest sites can limit the numbers of some hole-nesting birds such as Pied Flycatcher, which has spectacularly increased in many woods following the provision of nest boxes. Nest holes vary in quality and there can be great competition for the best sites (Chapter 3).

Dead wood is an important source of invertebrate food for woodpeckers, particularly in winter. A huge number of invertebrates depend on dead wood and dying wood. Many of these creatures live within dead wood for just a part of their life cycle; therefore, they form a potential food resource for many insectivorous birds, not just woodpeckers. It is possible that the amount of dead wood is of wide importance to woodland bird communities for, in a study of Swedish woods, Nilsson (1979b) found that the overall density and richness of birds was closely associated with the amount of standing dead timber (snags). Others have found relationships between snags and densities of hole-nesting birds (e.g. Haila *et al.*, 1987).

In conclusion, there is ample evidence that bird communities are richer in woodland habitats that are structurally diverse. Woods with a variety of growth stages will tend to be richer in bird species than uniform woods. Stands with a dense shrub layer, and rich in microhabitats such as gaps and snags, will tend to have higher bird densities than stands lacking these features.

Fig. 4.7. An open stand of birch woodland at the Muir of Dinnet National Nature Reserve, Aberdeenshire. A patch of regeneration can be discerned in the mid-distance, but otherwise the wood is devoid of a shrub layer. Derek Ratcliffe.

Tree species

The idea that the bird community of a wood should in some way be fashioned by the tree species composition, or floristics, is an appealing one. Tree species differ in the foods they offer birds (insects, seeds, fruits, leaves), they also vary in the amount of cover and nest sites they offer. One might expect, therefore, that bird communities would

differ in woods dominated by different tree species and be richer in woods that contain more tree species. Yet, unlike the effects of structure, convincing evidence has been rather slow to emerge, though it is now accumulating. There are two main reasons for this. First, few species of birds are dependent on one, or even a few, species of trees. Crossbill is the clearest exception, being adapted to a diet of pine seeds. Secondly, many of the interactions between tree composition and bird communities may be quite subtle and potentially easily masked by those of woodland structure. Relationships between bird communities and tree species composition are most likely to be detected when habitats of similar structure are compared (Fuller & Henderson, 1992; Fuller, 1994).

Traditionally, the description of woodland bird communities has been along the lines of comparing birds in oakwoods, beechwoods,

Fig. 4.8. Birch woodland with a dense understorey of juniper and considerable amounts of dead wood, Spey Valley, Invernessshire. This habitat offers a much greater variety of nest sites and feeding sites for birds than the birch woodland shown in Fig. 4.7. Derek Ratcliffe.

ashwoods and so on. This approach is of limited value because any observed differences could be due to a large number of factors. The woods have often differed in their location, soil types, altitudes and growth stages. Meaningful comparisons can be made only on sites that are close together and similar in all respects other than their tree composition. Despite this, there are fundamental differences between the bird communities of coniferous and broadleaved woods, though even here there is much to be learnt. Most species will occur in both woodland types but a few are confined to one or the other (Lack & Venables, 1939).

The following are largely confined to broadleaved woods in Britain: Lesser Spotted Woodpecker, Pied Flycatcher, Nuthatch, Marsh Tit, Wood Warbler, Nightingale and Hawfinch. Those largely confined to coniferous woods in Britain are: Capercaillie, Firecrest, Crested Tit, Siskin and Crossbill. There are many species that are generally more common in one type of woodland than the other, for example Coal Tit and Goldcrest in coniferous, and Blue Tit and Great Tit in broadleaved. In North America, conifers hold relatively few species and low densities of birds compared with broadleaves (James & Rathbun, 1981; James & Wamer, 1982). The same pattern has been described in Białowieża Forest, Poland (Tomiałojć & Wesołowski, 1990). In Finland, it has long been known that birch holds greater densities of birds than spruce, which in turn carries more birds than pine; these differences remain evident even when soil type is taken into account (von Haartman, 1971). One would expect similar patterns in Britain though, surprisingly, there have been few convincing studies. In the south of Wales, mature conifers were found to carry lower numbers of breeding bird species than mature broadleaves (Adams & Edington, 1973). In Scotland somewhat contradictory findings have emerged, for Newton & Moss (1977) working on Speyside reported higher densities of birds in birch than in pine, while in Deeside the reverse situation was found, though oak carried the greatest densities (French *et al.*, 1986). There is a clear need for well designed comparisons of bird communities in British broadleaved and coniferous woods. It would be especially interesting to know the effects of mixtures of broadleaves and conifers on bird communities. It is quite possible that mixed woods are the richest of all.

Insectivorous birds show preferences for feeding in certain species of trees (Ulfstrand, 1975; Holmes & Robinson, 1981; Morrison *et al.*, 1985; Peck, 1989). In a mixed plantation containing 13 tree species in northern England, there was an overall pattern of birds strongly preferring to feed in sycamore but avoiding beech (Peck, 1989). Nonetheless, there were substantial differences between bird species in their preferred tree species. Preferences for particular trees are likely to be underpinned by variations in the amount and types of insect food they provide. The morphology of the bird species themselves is also important, for some birds are better adapted to feed in certain types of tree foliage than in others. This leads to the prediction that woods with more tree species should support more bird species and this was indeed what Peck found. Among compartments of similar structure, she reported higher numbers of bird species, and higher densities of birds, in those woods containing the most tree species. There is still much to be learnt about the way in which birds use different tree species and it would be valuable to repeat this type of work elsewhere to assess the generality of this finding.

Although food is the main mechanism by which tree species influence bird communities, nest sites may be important for some birds. Trees differ greatly in their structure – in terms of foliage density, availability of holes and dead wood. This will certainly affect the suitability of different trees as nesting sites. In mixed conifer plantations, spruces offer better nesting sites for foliage-nesting songbirds

Fig. 4.9. Very little is known about the ecology of the Hawfinch in Britain, partly because nowhere is it numerous, and partly because it is extremely difficult to observe. It is confined to broadleaved woodland and is reputed to have an affinity with hornbeam for it eats large quantities of the seeds of this tree. Chris Rose.

than firs, pines and larch (Mackenzie, 1945). This preference for spruce is probably linked to the greater density of its foliage.

I shall conclude this section by briefly discussing the issue of invasive tree and shrub species. Sycamore and rhododendron are widely regarded as conservation problems, both having the capacity to invade woodland and spread very rapidly. Is invasion by these species likely to lead to major changes in woodland bird communities at a local level? For sycamore, there is no clear evidence that this is the case, though the issue has not been examined critically. Indeed, sycamore appears to be a preferred tree for feeding insectivorous birds because it supports a high biomass of insects, mainly aphids (Peck, 1989). It is quite possible, therefore, that the presence of *some* sycamore in a wood is beneficial to birds. The development of dense rhododendron probably does lead to major changes in bird communities. Where it invades woodland with a sparse understorey, it is likely to reduce the suitability of the habitat for species such as Tree Pipit, Wood Warbler and Pied Flycatcher, though more nest sites for thrushes may become available. Batten (1976) suggested that infestation by rhododendron reduced the diversity of bird life in Irish oakwoods. While rhododendron may not be good news as far as breeding birds are concerned, it should be recognised that it can offer a roost site for finches and thrushes in winter.

Social considerations

The distribution of some bird species, both within and among woods, may be influenced by the presence of other individuals of the same species in the following ways. Established birds may be used as a means of identifying suitable habitat (Reed & Dobson, 1993). Is it possible, for instance, that the absence of Nightingales from much apparently suitable scrub and woodland in England (Chapter 6) is a consequence of this type of behaviour?

There may be reproductive advantages in living close together. This has been suggested for the Wood Warbler by Herremans (1993) who found that territorial males were clustered in his Belgian study areas. The locations of these clusters varied from year to year, and they did not coincide with areas of caterpillar outbreaks, which would have offered a rich food supply, or with any obvious attribute of habitat structure. Females were scarcer than males and Herremans suggested that the formation of territory clusters may have been a strategy adopted by males to increase their chances of attracting a mate.

Demographic considerations

In the highly fragmented and unnatural landscapes of Europe there is likely to be considerable flux in bird populations between individual woods and the surrounding countryside. Even our largest tracts cannot be considered in isolation from events outside them. Some woods, or parts of woods, possibly act as *population sources* in that they produce more birds than can be accommodated by the suitable habitat there. The types of factors discussed earlier in this chapter would determine whether the habitat is of sufficiently high quality to act as a source. Birds may emigrate from such places to settle in other patches of woodland, some of which are perhaps of poorer quality and where insufficient birds are reared to maintain the population (in other words the death rate exceeds the birth rate). Here survival of the population depends on immigration of birds and such places could be regarded as *population sinks*. These ideas hinge on the abilities of species to move from one wood to another. It is unlikely that any species of British woodland bird is so tied to its birth site as to be incapable of colonising new habitat, but there is probably much variation in the relative dispersal abilities of different species. As discussed above, a small number of sedentary species may be unlikely to move readily between isolated patches of habitat.

Few species have been studied at an appropriate scale, for a sufficient period of time, and with adequate detail, to determine whether real populations behave in this way, effectively flowing between

Fig. 4.10. The Nightingale is confined to scrub and broadleaved woodland where there is dense, almost impenetrable, foliage up to some 2–3 m above ground. Breeding sites can soon become unsuitable for the bird unless the vegetation is managed. Declines in the British population of Nightingales in recent decades may partly be a consequence of habitat deterioration, for example the cessation of coppicing may have played a part, or perhaps the problem lies more on the wintering grounds. Much apparently suitable habitat is not used by the species. Is this simply because there are insufficient birds available to colonise it, or is the problem more complicated? It is possible that individual Nightingales are very faithful to those places where they have bred previously, or where they hatched. Alternatively, perhaps the birds use established territorial Nightingales to identify potential habitat. In either case, the birds would be unable to colonise new habitat readily. Chris Rose.

patches of habitat. One important exception, however, is the Sparrowhawk which has been intensively studied in conifer forests in southern Scotland by Ian Newton (1991). He has been able to identify certain places as high-grade habitat where breeding success was high and occupancy relatively constant. Breeding productivity at these places was more than sufficient to compensate for annual mortality. Low-grade places, however, produced fewer young, and occupation of these places was maintained only by immigration. For the study area as a whole the population was stable, with the production of young balancing the deaths. Therefore, for this species there was evidence of spatial differences in habitat quality, with some places acting as sinks and others as sources.

The population size of a species can have a profound influence on its habitat use. In The Netherlands, populations of the Chaffinch are more stable in mixed woods than in pine woods (Glas, 1960). Birds prefer mixed woods and only when population levels are high does the pine hold many birds. These include a disproportionately high number of first year birds which have been unable to obtain a territory in the mixed woodland. Hence, the pine population can be thought of as overspill. Another example is given by the Wren, which appears to prefer the edges of some English woods to the interior (R.J. Fuller, unpublished). When populations have been reduced by hard winter weather, the territories within these woods are located mainly at the edge. As the number of Wrens increases, so the woodland interior becomes colonised. Presumably the breeding success of the birds in their preferred habitats – mixed woods for the Chaffinch and woodland edges for the Wren – is higher than in those habitats which are occupied only when the population level is high, but this has not been tested.

This chapter has been largely concerned with explaining why it is that birds vary so much in abundance from one wood, or part of a wood, to another. It has finished, however, with the message that while the density of a species will often be a good indicator of habitat quality, this is not always the case. The best habitats (i.e. ones that are likely to be population sources) will be those where breeding success, or overwintering survival, is highest. Though we know a lot about the numbers of birds in woodland, unfortunately we know very little about their breeding performance in different types of woodland.

5

Scrub

It is logical to treat scrub before woodland because it is usually a transient vegetation, eventually giving way to woodland of one kind or another. Scrub deserves recognition as an important bird habitat in its own right. In the 'wrong place', scrub can be a real problem, for example where it is invading open heaths or chalk grassland rich in flowers. In these cases, scarce types of vegetation can be rapidly overwhelmed unless the scrub is held back. Elsewhere, however, the establishment of scrub can be a real gain from a conservation viewpoint. The insect and bird life associated with scrub can be extremely rich; the density of birds breeding in thick scrub can be far higher than that found in much mature woodland. There is considerable overlap in the species composition of bird communities living in scrub and those living in mature woodland. Nonetheless, scrub offers a range of habitats and food resources that are largely unavailable to birds in mature woodland. Young-growth in plantations, but perhaps more so in coppice, can support a similar bird life in summer to that of scrub but seldom offers such large quantities of fruit for birds in autumn. Dense scrub is also an important roosting habitat for thrushes, finches, buntings and Starlings.

The variety of scrub types in Britain is enormous and there is much yet to be learnt about their bird life. Scrub habitats often have much bracken and bramble within them. While scrub and bramble can form important nesting, feeding and roosting sites for birds, this is not the case for bracken which is used very little by birds (Beven, 1964). In the lowlands, the commonest scrub is predominantly hawthorn, while in the Scottish uplands it is generally birch, though in Wales the slopes on the fringes of the hill land (the *ffridd*) are typically covered by bracken and thorn bushes. Many commons in the south of Britain are a mixture of grassland, bracken and thorn scrub. Gorse, birch and pine scrub are widespread on lowland heaths and, in moderation, add considerably to the diversity of heathland bird life. Downland and other calcareous soils carry a rich mixture of shrub species offering a wide range of insects and berries for birds to eat. On heavy lowland soils, thickets of blackthorn can rapidly establish themselves and provide dense cover for many birds. British wetlands also hold several distinctive types of scrub, ranging from pure willow to dense thickets of buckthorn and alder buckthorn. Arguably, the most important scrub sites for birds occur on the coast, where huge numbers of birds depend on them for food and shelter during migration. These include intricate mosaics of bracken, bramble and scrub on the upper slopes of seacliffs. In south-west Britain, sheltered coastal valleys also harbour varied scrub habitats that accumulate vast numbers of migrant birds and also act as a winter refuge for many passerines. The large stands of sea buckthorn and other shrubs, characteristic of coastal dunes in eastern England, represent an entirely different scrub environment. This scrub is important as a feeding site for migrating warblers, but it can also hold high breeding densities of Sedge Warbler, Whitethroat, Dunnock, Linnet, Redpoll and Reed Bunting (Morgan, 1978).

Virtually any piece of lowland countryside where there is no cultivation, grazing or cutting of the vegetation will sprout bushes, most

Fig. 5.1. Downland scrub on the Chiltern escarpment at Aston Rowant National Nature Reserve, Oxfordshire. The scrub consists of patches at different stages of growth, each of which are used by a different suite of bird species. Peter Wakely, English Nature.

commonly hawthorn, blackthorn or elder. If left unchecked, these bushes may gradually form an impenetrable thicket. Tree saplings growing up among the bushes eventually outgrow the shrubs and the vegetation starts to take on the appearance of young woodland. Usually the only way to prevent this progression to woodland is to cut or graze the scrub but there are situations where scrub can persist without such intervention. On storm-swept coastal cliffs, scrub can probably exist indefinitely, often in a mosaic with bramble, bracken and heath. It is also likely that the scrub phase lasts for a very long period on many dunes. Blackthorn forms extremely dense stands within which regeneration of other shrubs and trees can be very limited. The stand can be perpetuated by fresh suckers which spring up to take the place of old blackthorn stems when they topple and die. Some fenland scrub has a similar capacity to stave off woodland for many years. At the upper limit of tree growth on Britain's mountains, woodland would have naturally given way to a zone of permanent scrub of willows, birch and juniper, lying beneath the open high tops. However, in the face of ubiquitous grazing, examples of this sub-montane scrub are few and far between.

It is not my intention to present an account of the bird life associated with each of these types of scrub. The following account of birds in downland scrub gives an outline of the dynamic characteristics of breeding birds in scrub, which probably applies to much scrub in Britain. The aim of this chapter is to summarise how bird populations respond to the development of scrub, to describe how birds use scrub outside the breeding season, and to discuss various approaches to the management of scrub. The vast tracts of scrub in Mediterranean Europe hold a distinctive avifauna which is also briefly described here because it makes an interesting comparison with temperate scrub.

Responses of breeding birds to scrub growth: a downland example

Scrub has spread across downland in southern England in recent decades as traditional sheep grazing has declined and myxomatosis has greatly reduced rabbit populations. Not surprisingly, these events have transformed the bird communities found on these chalk hills. The available habitat for birds of the open downs, such as Lapwing, Stone-curlew and Skylark, has diminished. Overall, however, the numbers of bird species breeding on many downland areas have increased as birds requiring scrub and trees have been able to establish themselves. This is also true for other types of land where scrub is encroaching (Melluish, 1969).

In the early 1980s, I studied the distribution of breeding birds on the downland escarpment of the Chiltern Hills in Buckinghamshire and Hertfordshire (Fuller, 1987). Several patches of chalk grassland remain free of bushes, but much of the escarpment has long since become smothered by scrub (Fig. 5.1). In some areas hawthorn is the dominant shrub, but elsewhere a more intimate mixture of shrubs occurs including privet, wayfaring tree, dogwood, buckthorn and juniper. The scrub exists as patches at all stages of growth from scattered small bushes to closed-canopy bushes more than 6 m tall. Beneath the oldest scrub, the ground is often bare and there is little foliage close to the ground. Because the vegetation of the escarpment is so varied it was possible to examine how the bird communities changed across the successional gradient from open grassland to mature scrub (Fig. 5.2).

The numbers of breeding species and the total numbers of individual birds increased strongly across this gradient in very similar fashions. Locations with 10% or less scrub cover typically held no more than five species. Nearly all points where the scrub cover was 50% or more held at least five species and often more than ten. The numbers of species and individuals increased almost linearly up to 75% scrub cover, but then levelled off, before decreasing slightly in the most closed scrub. The composition of the bird community changed most rapidly in the earliest stages of scrub development. An increase of scrub cover from 5% to 25%, made a much larger difference to the list of breeding species than did an increase from 35% to 60% or from 60% to 80%. In other words, the addition of even a modest amount of scrub had a substantial impact on the bird life. The reason for this is twofold. First, species associated with grassland, for example Skylark, disappeared quickly once the scrub invaded. Secondly, a large number of species live in quite open scrub, even those which reach their highest densities in older, denser scrub.

Individual species differed greatly in their distribution across the scrub gradient (Fig. 5.2). No species was evenly distributed across all stages; most reached maximum numbers at a particular stage of scrub growth. Skylark and Meadow Pipit were the two species most characteristic of open grassland but, whereas the lark was strongly intolerant of scrub encroachment, the pipit only declined when scrub cover exceeded some 25%. Tree Pipit completely avoided open grassland and reached peak numbers at the stage when Meadow Pipit was well into its decline. Linnet, Whitethroat and Yellowhammer were similar to Tree Pipit in that they were confined to open-canopy scrub. Several other species also first appeared at a very early stage of scrub growth but persisted for much longer. Willow Warbler, for example, was the dominant bird throughout much of the scrub development but it declined after canopy-closure. Garden Warbler and Dunnock also peaked at about the canopy-closure phase. These three species probably require a combination of a high density of bushes and reasonably low, thick foliage which tends to shade out rapidly when the canopy closes. Blackbird, Song Thrush, Robin, Chaffinch and Wren also built up to maximum numbers at about canopy-closure but they did not subsequently decline. Hole-nesters were scarce and largely confined

Fig. 5.2. Abundance of breeding birds on the escarpment of the Chiltern Hills in relation to scrub growth. Based on point counts conducted in 1980 and 1981 (Fuller, 1987). The index of abundance is derived from numbers of birds counted within a 50 m radius at more than 90 locations.

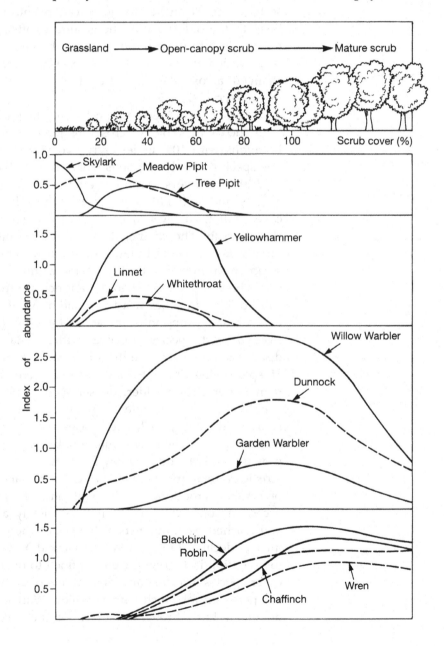

to closed-canopy scrub. Large elder bushes often provide many of the nest sites for hole-nesters in scrub. This turnover in species is very similar to that which occurs in coppiced woodland (Chapter 6). Though this example is drawn from downland in southern England, these changes in bird communities are likely to be similar to those that occur in many other types of scrub succession.

Scrub can be a particularly rich breeding habitat for summer visitors. Where a wide range of growth stages is present, downland scrub can support seven species of warblers. There is considerable overlap in their habitat requirements but they can be broadly summarised as follows. Grasshopper Warbler and Whitethroat prefer the most open scrub, where the bushes are interspersed with rank grass. Willow Warbler prefers denser scrub; it can be the most abundant species when the scrub has reached 40% canopy cover or more. Garden Warbler and Lesser Whitethroat select patches of dense bushes, though the latter species probably prefers taller growth. Blackcap does not usually colonise scrub before the closed-canopy stage. Chiffchaff is a less typical breeding warbler of scrub; its occurrence seems to depend on the presence of some tall trees. Sedge Warbler will also breed in low scrub, especially at wet sites, but rarely, if ever, in downland scrub. The overall densities of warblers breeding in scrub are probably only matched by those found in the early and middle stages of growth of some coppiced woods. Other species of migrants that will nest in dense scrub are Turtle Dove and Nightingale, the latter being very similar to Garden Warbler in its selection of breeding sites. Until some 20 years ago another migrant, the Red-backed Shrike,

Fig. 5.3. The range of the Turtle Dove in Britain has contracted substantially towards eastern England in recent decades. Dense scrub is among its most important nesting habitats. Chris Rose.

was also characteristic of much scrub in southern England. It held territories in areas with scattered bushes or where dense scrub abutted more open patches. The decline of the shrike to virtual extinction in Britain has still not been satisfactorily explained. Suitable habitat remains for the species at many of its former sites. It seems that the reason must lie either with some factor operating in its African winter range, or with a reduction in the food supply here in Britain. The large invertebrates which it eats may have declined in response to some subtle climatic change, or to agricultural intensification.

Scrub on lowland heaths

Heathland is locally distributed on impoverished gravels, sands and acid soils. It is not natural, but rather a product of ancient woodland clearances. The main concentrations are in Breckland, the Suffolk Sandlings, Surrey and north-east Hampshire, the New Forest and south Dorset, though scattered outliers exist as far apart as western Cornwall and Lincolnshire. The vegetation of dry heathland is typified by low shrubs usually dominated by ling, often with bell heather and dwarf gorses, which give way to a different mixture of plants in wetter areas. Grass heaths, often interspersed with heather, are a feature of Breckland. Patches of bracken and dense gorse scrub have probably always been an integral part of heathland vegetation.

Few species of birds use those parts of heaths that consist purely of the low dwarf shrub vegetation. Meadow Pipit is by far the commonest, though the much rarer Woodlark may be found on some southern heaths where there are patches of bare ground. In recent decades, Woodlarks have greatly declined on heaths in eastern England, probably because grass and scrub have encroached on their open feeding areas. The richness of the heathland bird community increases wherever patches of gorse scrub occur. This was clearly demonstrated in a study on Dorset heathland in which the distributions of the four commonest insectivorous songbirds were mapped throughout the year in relation to the distribution of gorse (Bibby, 1978). The species were Stonechat, Dartford Warbler, Wren and Goldcrest, the latter two mainly used the habitat in winter. Gorse probably offered better winter food supplies for insectivorous birds such as Wrens and Goldcrests than many deciduous shrubs. Gorse covered only 2% of the study area, yet it was extremely important to each of these species except Stonechat which fed relatively often on the open heath wherever hunting perches were available. The distributions of Wren, Goldcrest and Dartford Warbler closely matched that of the gorse. Furthermore, the biomass of birds per unit area was far greater in the gorse than it was in the heather and this difference was especially marked in winter. Throughout the year, the gorse offered a richer food supply for birds than did the heather. Dartford Warblers nesting in rank heather would fly considerable distances to patches of gorse from which they collected food (Bibby, 1979*b*).

Lowland heath has become one of the most threatened habitats in Britain (Cadbury, 1989). The total area of open lowland heath has been massively reduced this century by habitat destruction. Heathland seems to have come under pressure from every direction, with losses from urban and recreational development, road building, agriculture and forestry. Natural processes are, however, one of the major agents of change on heaths. Heathland can only persist where the vegetation is controlled by cutting, grazing or burning. In the absence of these controls, heath moves inexorably towards woodland.

One of the traditional uses of heaths was as grazing for sheep and cattle, but this has long since declined in all regions. In the New Forest grazing does persist, though mainly by ponies. The lack of grazing has triggered widespread changes in vegetation. The seeds of birch and pine are readily dispersed and these trees are rapidly smothering dwarf shrub heath at many sites. Without large-scale intervention much lowland heath will change to woodland very rapidly. The speed of the transformation can be illustrated by two examples from different parts of England. Surveys of heathland in Dorset in 1978 and 1987, found that in just 9 years the area of this invasive scrub had increased by 15% (Webb, 1990). Analysis of aerial photographs of three Breckland heaths has revealed a massive spread of scrub and woodland since the late 1940s (Marrs *et al.*, 1986). Over approximately 35 years, the cover of these sites by scrub and woodland has risen as follows: 19% to 52%, 7% to 27%, 0% to 50%. Which tree species colonise depends largely on the proximity of seed sources. Heaths with existing birch woodland will tend to develop birch scrub, while close to pine plantations the spread of pine is a particular problem (Fig. 5.4). Bracken encroachment does not seem to be such a severe problem because the bracken spreads from existing patches and can be relatively easily kept in check at the invading front (Marrs *et al.*, 1986).

The bird communities of invasive woodland on heathland may well carry more species and higher densities of breeding birds than did the heathland that it replaced, but this is not an argument in its favour. Most of the bird species colonising the birch and pine scrub are widely distributed and could not be regarded as any compensation for the loss of the highly specialised and unique communities of plants and animals living on open heathland. Moreover, the scrub on heathland does not have the same biological interest as that on down-land, for example, where the scrub is usually composed of many more species and probably attracts a much wider variety of invertebrates. Densities of birds in downland scrub are likely to be much higher than in heathland scrub at a similar successional stage. While a fundamental tenet of conservation is the maintenance and promotion of biological diversity, the conservation of certain species-poor habitats, such as heathland, must be a high priority. This is not the paradox it may seem because heaths make an important contribution to biodiversity at a regional scale, simply because several species depend on them. Limited areas of trees and scrub can do much to diversify the bird life of heathland, benefiting not just common species, but birds such as Nightjar and Tree Pipit which are often associated with scattered trees, or the edge of woodland where it abuts heath. The problem is one of preventing the trees from overwhelming the dwarf shrub vegetation and I return to this subject in the section on scrub management.

Fires are a natural part of heathland ecology but if they occur too frequently they are likely to cause severe reductions to the populations of rare heathland animals, and perhaps aid the spread of bracken and birch. The short-term effects of a heathland fire can be devastating but bird populations have the capacity to recover quite quickly. This is shown by a fire in 1976 on a Sussex heath which consisted of a complex mosaic of bracken, heather and gorse (Hughes & Griffiths, 1983). The fire destroyed the vegetation covering 90% of the site.

Fig. 5.4. The spread of Scots pine at Lakenheath Warren, Suffolk, between 1971 and 1984. From Marrs *et al*. (1986), with permission from Elsevier Science Ltd.

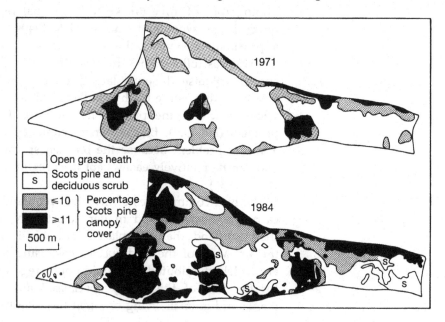

Counts of ten bird species breeding in the open habitats and in the scrub showed a 60% decline in numbers in the year after the fire. Four years after the fire, the numbers of most species had returned to their pre-fire levels. The rapidity of this recovery was surprisingly fast considering that most of the species were dependent on the re-establishment of scrub.

Use of scrub by migrating and wintering birds

Scrub can provide a prolific source of berries for birds during autumn and early winter. The amount of berries in scrub is often far greater than in an equivalent area of mature woodland where berries are often found in any quantity only at the edges and in treefall gaps. The large variety of berry-bearing shrubs found in much downland scrub can attract spectacular gatherings of thrushes. The autumn influx of Song Thrushes, Blackbirds, Redwings and Fieldfares into scrub at a site on the South Downs is illustrated in Fig. 5.5. Here the main foods were elderberries and hawthorns (Leverton, 1986). In early autumn, small numbers of migrating Ring Ouzels also fed on elder-berries from scattered bushes. The numbers of Redwings and Field-fares were especially variable from one winter to another, perhaps in response to the size of the crop of hawthorn berries. Once the haws had been consumed, usually by the end of December, the thrushes abandoned the scrub, though some continued to roost there at night. Several species preferred gorse scrub for roosting, perhaps because it was evergreen.

Warblers also concentrate in scrub in order to feed on fruits in

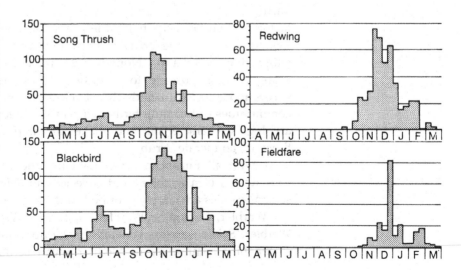

Fig. 5.5. The seasonal occurrence of thrushes in downland scrub on the South Downs, Sussex. Total numbers of thrushes caught and ringed between 1975 and 1985 are shown for a site which was worked throughout this period. From Leverton (1986).

autumn. Fruits are a particularly important food for *Sylvia* warblers. Taking Blackcap as an example, large numbers build up at particular localities where there is a regular abundant supply of berries. Some of these accumulations are inland, such as one place on the Chiltern escarpment where the birds feed heavily on berries of the wayfaring tree. The largest concentrations, however, are probably on the coast. One of the most important sites is at Beachy Head in Sussex, where large numbers of migrating Blackcaps, probably mainly of British origin, feed especially on fruits of bramble and elder (Edgar, 1986). The importance of elderberries as an autumn food of *Sylvia* warblers has also been demonstrated on the east coast (Boddy, 1991).

There is evidence that wet scrub is a more important habitat for juvenile warblers during late summer than are dry habitats (W. Peach, personal communication). Large-scale bird-ringing programmes have shown that the ratios of juvenile to adult Blackcaps, Willow Warblers, Chiffchaffs and Robins caught in mist nets are consistently higher in wet than dry scrub. This may be related to the existence of a greater abundance of insect food, rather than of fruit.

Mediterranean scrub

The scrub habitats of the coastal and hill regions of the Mediterranean countries are more extensive than those found anywhere else in Europe. They occur from Iberia to Greece and embrace many types of shrubland differing in shrub species and structural form. There are specific names for the different types of shrublands, notably maquis and garigue (see glossary). This complex environment is partly a product of historical land-use, which has impoverished the soils, and partly of topography and climate. Fire and grazing are important factors acting to perpetuate these shrub communities, though the degradation of some areas is such that they may be incapable of supporting woodland. Nonetheless, in the south of France for example, it is possible to identify patches that represent different successional stages towards the development of holm oak forest. The communities of breeding birds living in the Mediterranean shrublands are highly characteristic and make an interesting comparison with those of temperate scrub.

As with all European scrub habitats, the Mediterranean scrub is extremely rich in warblers, but especially in *Sylvia* species. In Europe, several species are largely confined to these habitats, including Sardinian Warbler, Subalpine Warbler, Spectacled Warbler and Marmora's Warbler. *Sylvia* warblers generally make up a much higher proportion of the bird community in Mediterranean scrub than is the case in temperate scrub. In the south of France, *Sylvia* species can be the commonest birds in scrub between 1 m and 6 m tall (Blondel, 1981). Depending on the exact type of vegetation, the species concerned are Dartford Warbler, Sardinian Warbler and Subalpine Warbler. Nightingale also reaches extremely high densities and, in scrub of about 4 m tall, can be one of the two or three commonest birds. Another difference between Mediterranean and temperate scrub is that a higher proportion of warblers breeding in the south are residents. Of the five *Sylvia* species breeding in Britain, three are trans-Saharan migrants (Whitethroat, Lesser Whitethroat and Garden Warbler), one is mainly a medium-distance migrant (Blackcap), and one a resident (Dartford Warbler). Of the seven in the extreme south of France, three are trans-Saharan migrants (Orphean Warbler, Whitethroat, Subalpine Warbler), one is a short-distance migrant (Spectacled Warbler) and three are resident (Dartford Warbler, Blackcap, Sardinian Warbler). Interestingly, four of the typical birds of scrub in Britain – Garden Warbler, Whitethroat, Lesser Whitethroat and Willow Warbler – are rare or absent from Mediterranean scrub (Blondel, 1981).

Enormous numbers of passerine birds overwinter in the scrub and woodland habitats of the Mediterranean shrublands. These include both insectivorous and frugivorous species. Many of these birds originate from central Europe where normal winter conditions are too severe for their survival. Some species from western Europe, however, also overwinter in shrubby habitats around the Mediterranean. For example, substantial proportions of the British breeding populations of Stonechat, Blackcap and Chiffchaff, winter in southern Europe. Fruit is an important food for the populations of thrushes, Robins and Blackcaps overwintering in the Mediterranean region. Though fruit represents a huge food resource for these birds, the amount of berries varies considerably from year to year. However, at least in southern Spain, the size of the berry crop does not seem to determine the numbers of Robins and Blackcaps wintering in local areas of scrub (Herrera, 1988). Individual Robins and Blackcaps can show high fidelity to particular Mediterranean wintering areas, both within the winter period, and by returning to the same area in subsequent winters (Cuadrado, 1992).

The control and management of scrub

Controlling the spread of scrub is necessary where it threatens a valuable habitat such as heathland. As described above, the retention of a modest amount of scrub on heathland will add much to the ornithological interest of the site. If scrub development is allowed to proceed too far, however, the special birds of heathland, such as Dartford Warbler and Nightjar, will eventually disappear. Heathland management is beyond the scope of this book and is considered in detail elsewhere (Andrews, 1990; Dolman & Land, 1994). Nonetheless, it is worth briefly considering some of the approaches and problems of scrub control here. Where already established, the usual approach is to cut down individual trees, though herbicide treatment of birch may be appropriate in certain circumstances (Marrs, 1985). Scrub cutting has been successfully employed by the RSPB on its Suffolk heathland where it has been invaded by birch and pine (Burgess *et al.*, 1990). The work involved cutting glades in the woodland and increasing the amount of heath-woodland edge by felling. Other management included the creation of shelterbelts, by natural regeneration on open heath, and the clearance of small bare patches beneath trees. Over a 12 year period since 1978 when the management was started, the Nightjar population rose five-fold. Where it is necessary to control bracken, this can be achieved effectively with the herbicide Asulam (Andrews, 1990). In the absence of burning or grazing, cutting will need to be repeated at intervals. On large heaths, burning is sometimes used as a management technique to encourage a variety of age classes of heather, but obviously this can be a risky business.

Management of scrub as a habitat in its own right is highly desirable where this does not seriously conflict with other conservation aims. Scrub is a valuable habitat for wildlife especially insects and birds, but it does need to be managed to retain its special interest. Extensive areas of old 'leggy' scrub are of relatively little value for birds compared with scrub which has recently closed canopy or where the bushes are interspersed with grassland. The birds that breed in scrub differ greatly in their preferred stages of scrub growth (Fig. 5.2). Therefore, to sustain the richest communities of scrub birds it is necessary to maintain a variety of scrub-growth stages, from recently cut to closed-canopy. This can be done by rotational cutting, which has proved a successful means of enhancing the bird density in willow scrub (Wilson, 1978). The exact rotation time will depend on the species and growth rate of the scrub but, as a rough guide, cutting should be carried out some 5 to 8 years after the canopy has closed. The cutting of small patches would create a complex habitat structure with many edges which may be beneficial to both insects and birds. There have been no systematic studies of the responses of birds to scrub-cutting comparable to those of coppiced woodland where changes in the bird life have been documented in relation to the felling cycle (Chapter 6).

Large mammals offer great scope for both controlling and innovatively managing scrub, especially on nature reserves (Fuller & Peterken, 1994). From a long-term conservation viewpoint, it seems desirable to re-instate grazing on heathland as a means of controlling scrub, and of creating interesting mosaics of vegetation which include

Fig. 5.6. Birch scrub encroaching on open heathland at Thursley Common National Nature Reserve, Surrey. In the absence of heathland management such scrub can rapidly smother areas of heather. Peter Wakely, English Nature.

some scrub. Grazing on heathland may not be commercially viable, so the exercise is likely to need subsidising. It would be important to guard against nutrient import through supplementary feeding. Furthermore, rather little is known about the effects on the vegetation of different rates of stocking, so careful trial and error would be needed to avoid damage by trampling or overgrazing (ideally experiments on the effects of grazing on heaths are needed but time is too short to wait for refined answers). Despite these difficulties, grazing has long been a central component of heathland ecology and, in the long run, may prove a more sustainable approach than scrub-cutting on important heathland areas. In other habitats such as dunes, fens and downland, low grazing pressure by horses, sheep or cattle may allow scrub to persist alongside grassland in a sort of dynamic equilibrium. It should be possible to manipulate grazing to produce intricate, patchy scrub habitats supporting a wide range of birds throughout the year. The resulting vegetation would have a more natural appearance than that produced by cutting.

6

Broadleaved high forest and coppice

Nearly all managed broadleaved woods in lowland Britain are some form of high forest or coppice. In addition, many unmanaged woods were once coppices that have fallen into disuse because they could no longer be worked economically. It is likely that eventually many derelict coppices will be converted to high forest. High forest and coppice systems produce woods of very different structure and, accordingly, their bird communities are strikingly different. This chapter describes and compares the main features of the bird life of high forest and coppice, and the changes that occur as the woods develop through the felling cycle. The emphasis is on lowland woods. There are many variations on the basic themes of high forest and coppice management (Chapter 1), which can have profound implications for birds and other wildlife. The subject of woodland management and its effects on birds is, therefore, dealt with in some detail. Grazing is an increasingly important influence on the structure of much woodland in lowland Britain but discussion of this issue is deferred to the following chapter.

An outline of high forest bird communities

Within broadleaved woods managed by clear-felling or shelterwood, there are striking differences in the bird communities found at the various stages of growth. Similar changes take place in coniferous plantations between planting and felling (Chapter 8). Some of the changes in the bird community of a lowland French oak forest are illustrated in Fig. 6.1. I have chosen a French example partly because broadleaved high forest bird communities have been studied in greater detail there, and partly because the stronger tradition of high forest in France has maintained far more extensive and impressive stands than in Britain. Similar patterns are evident in the bird life of British oakwoods, though Wood Warblers are very patchily distributed in our lowland woods, even in the older stands. These large changes in the bird life are driven by the transformation of the woodland structure as the trees grow, and the canopy gradually closes, shading out much of the low vegetation.

There appear to be three broad phases in the bird communities (Ferry & Frochot, 1970). In the earliest years (establishment and early-thicket stages), warblers are the dominant group, though other common species may be Tree Pipit, Dunnock and Yellowhammer. The mid- and late-thicket stages form an intermediate phase, in which a moderate density of warblers persists but numbers of Robins and large thrushes increase, while hole-nesters start to make an appearance. The post-thicket stands represent the third phase in which five species appear to be consistently among the commonest breeders, at least in lowland British oakwoods: Wren, Robin, Blue Tit, Great Tit

Fig. 6.1. Changes in the densities (pairs/10 ha) of breeding birds in three genera in relation to the age of stands within French oak high forest. From Ferry & Frochot (1990).

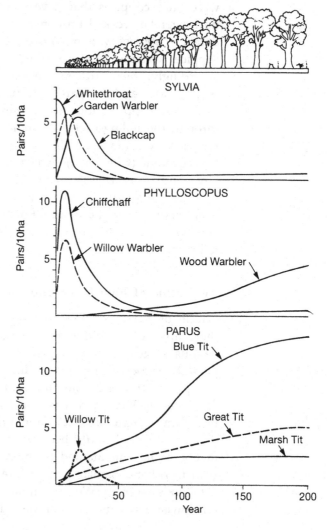

and Chaffinch. These carry the greatest densities of hole- and crevice-nesters: Stock Dove, Jackdaw, woodpeckers, Redstart, flycatchers, Starling, tits, Treecreeper and Nuthatch. The populations of some of these hole-nesting birds continue to increase well beyond the thicket stage. Densities of Blue Tits, for example, may continue to rise in stands older than 100 years (Fig. 6.1). The only non-hole-nesters that are specialists of old stands are Honey Buzzard and Hawfinch. Among the hole-nesters in the French forests, Willow Tit is an exception because it is associated with fairly young stands, presumably due to the presence of birch trunks which are effectively weeded out as the trees grow.

Results from a study in the Forest of Dean, in stands ranging from 32 to 174 years of age, demonstrate some of the major factors influencing the nature of oakwood bird communities (Smith *et al.*, 1985). These woods lie at the boundary of lowland and upland Britain and their bird communities show characteristics of both regions. Not surprisingly, the strongest factor was the age of the trees. The overall density and diversity of birds increased with the age of the trees. Several hole-nesting species, in particular, were more or less confined to the oldest stands, though densities of most of the commoner birds were highest in the oldest stands. The species composition was also related to the structure of the shrub layer. Some species, notably Tree Pipit, avoided areas with a dense shrub layer while others, including Marsh Tit and Blackcap, preferred places where the shrub layer was better developed.

Breeding bird communities in mature oak and beechwoods are better documented than those in other types of lowland broadleaved high forest (Yapp, 1962; Simms, 1971). They illustrate how different tree species can have an indirect effect on birds through the food supply and the structure of the habitat. Chaffinch is generally the commonest breeding bird in both types of wood. Several species, most strikingly Blue Tit and Nuthatch, prefer oakwoods. Blue Tits live at higher densities in oak, whereas Nuthatches can be equally abundant in both habitats, but their breeding success is higher in oak (Nilsson, 1976). In Britain, there are no species that clearly prefer breeding in beech to oak, though in some lowland areas Wood Warblers are associated mainly with beechwoods, presumably because they favour woodland with sparse undergrowth. Overall densities and numbers of species are usually considered to be higher in oak than in beech, though this depends to some extent on the structure. However, in mixed oak–beech woodland in northern France, Smith (1992) could find no relationship between the diversity and overall density of breeding birds and the ratio of oak to beech trees. Much mature beech exists as stands with hardly any field or shrub layers (Fig. 6.2). These stands are virtually devoid of birds that depend on these lower strata of vegetation. Where secondary growth does exist within beech, as on slightly richer soils, or where the canopy has been broken up by group-felling, the bird communities are appreciably richer, though usually less so than in oak of comparable structure. Perhaps the main explanation for the relative richness of bird life in oak is that this tree supports far larger populations of defoliating caterpillars than beech. This superabundant food supply is exploited by a wide range of birds, and not only by those breeding within the wood. Birds breeding outside, for instance Blackbirds and Starlings, will come into oakwoods in early summer to feed on the caterpillars (Beven, 1976). In the case of Starlings many of these birds are juveniles. Unlike beech, the understorey of oak woodland often includes much hazel and hawthorn which is also rich in insects (Fig. 6.3).

Rather little is known about high forest bird communities outside the breeding season. Several 'resident' species abandon some broad-leaved woods in winter for other habitats. For example, Beven (1976) found that Robins, Dunnocks and Chaffinches largely deserted a

Fig. 6.2. Beech woodland in the Chilterns at Bradenham, Buckinghamshire. This wood is managed by group-felling which creates a mosaic of patches at different stages of growth. The old patches are largely devoid of any undergrowth in contrast with the dense low vegetation in the recently felled areas. Peter Wakely, English Nature.

Surrey oakwood in the autumn. Oak and beech both produce large seeds which are eaten by birds. Acorns are an important food of Jays and Woodpigeons, but a wider range of species eats beech seed. Large numbers of birds can be drawn into beechwoods in 'mast years', including Woodpigeons, Jays, several tits, Chaffinches and Bramblings. In Swedish beech woodland, Nilsson (1985) found that Woodpigeon and Brambling were the most important predators in terms of the amounts of beech seed eaten. The population dynamics of some birds shows close links with the size of the beech crop. In years following a winter when beechmast is abundant, Great Tits

Fig. 6.3. Oak woodland with a well developed shrub layer in The King's Forest, Suffolk. Stands of this structure typically hold high densities of Nuthatches and other hole-nesting birds. Derek Ratcliffe.

breed at higher densities (Perrins, 1979). In autumns when beechmast is plentiful, Nuthatches hold smaller territories, their densities are higher and a larger population breeds the next spring (Enoksson & Nilsson, 1983; Nilsson, 1987). Though both these species feed extensively on beech seed, it is possible that the birds are really responding to other food sources that fluctuate in synchrony with beechmast; this seems likely in the case of Great Tit which, even outside the range of beech, fluctuates in parallel with the beech crop. Outside the range of beech, the Nuthatch may show fluctuations that correlate with the seed crop of other trees, such as hazel (Enoksson, 1990). Annual variation in the amount of seed produced by trees also affects densities of small mammals in woodland. The breeding success of Buzzards and Tawny Owls in woodland can be closely linked to such fluctuations in the availability of mammal prey (Southern, 1970; Tubbs & Tubbs, 1985).

Most conifers are harvested at less than 60 years of age but many broadleaves are felled well after 100 years of growth. From an ornithological viewpoint this is significant because, with increasing maturity, there are greater amounts of dead wood in the form of standing dead trees, dead limbs on living trees, and fallen timber. Nesting cavities tend also to be increasingly available for hole-nesting birds. These general patterns in availability of microhabitats will be modified by the tree species of which the wood is composed, and by practices such as clearing out dead wood and thinning. Nonetheless, broadleaved woods close to felling age tend to carry much higher densities of hole-nesting birds than pre-felling coniferous woods. This effect is not simply due to differences of habitat maturity; British broadleaved woods appear to be intrinsically more favourable to hole-nesting birds than conifers. Four of the seven specialists of broadleaves are hole-nesters, compared with one of the four conifer specialists (Chapter 4). Of the five species of *Parus* tits breeding in lowland Britain, the largest genus of hole-nesting birds, four prefer broadleaved woods (Marsh Tit, Willow Tit, Blue Tit, Great Tit) and one prefers conifers (Coal Tit).

High forest systems and thinning regimes

The basic options for managing broadleaved high forest are clear-felling, shelterwood, group-felling and selection (Chapter 1). Nowhere in Britain is it possible to make a valid comparison of the four systems, partly because shelterwood and selection are seldom practised here on an adequate scale. Nonetheless, some speculation about their bird communities seems reasonable, based on the contrasting types of habitat structure created by each system. In making this comparison I have assumed we are comparing four hypothetical woods, identical in every respect except for their management system, each wood being 100 ha and managed on a 100 year rotation. In reality, this simple situation is hardly likely to exist! Given sufficient information about the structural habitat needs of each bird species, it would be possible to model the responses of bird communities to the four systems when both woodland size and rotation length varied. The size of patches (coupes) could also be manipulated within the systems. But given our restricted knowledge, we are limited here to a general discussion.

Clear-felling and selection lie at opposite extremes of a continuum of woodland structure created by high forest management (Fig. 6.4). It simplifies matters, therefore, to start by comparing these two fundamentally different systems. In clear-felling, the wood consists of a coarse-grained patchwork of large blocks, each containing many trees of the same age. Our hypothetical clear-felled wood has a coupe or patch cut every 5 years giving a total of 20 patches, each of 5 ha, representing a wide range of growth stages and habitat structures. Selection systems, on the other hand, create a horizontal structure which appears far more uniform and constant, though in reality it is a fine-grained, dynamic patchwork carefully controlled through the removal and promotion of individual trees. The foliage profile of a selection wood is extremely complex – at any one point there will be

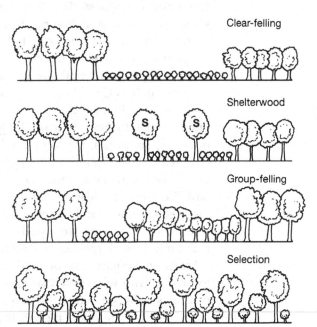

Fig. 6.4. Diagrammatic woodland structure under four systems of managing high forest. Seed trees in the shelterwood are denoted S. See Chapter 1 for more details.

several layers of foliage due to the mixture of age classes throughout the wood. In clear-felled woods, however, the foliage profile at any one place is unlikely to be so complex, even within the mature stands. Based on this knowledge of woodland structure, the following *predictions* can be made about the breeding bird communities associated with clear-felling and selection.

1 Bird species that need either open spaces, or extensive areas of bushy vegetation without a tall canopy above, will be favoured by clear-felling. Such species include Tree Pipit, Whitethroat and Yellowhammer.
2 Both systems provide suitable habitat for birds dependent on mature forest, but it is possible their numbers would be larger in the selection wood because mature trees are spread over the entire wood, whereas under clear-felling, mature trees will be absent from a large part of the wood.
3 Clear-felling, on the scale discussed here, creates a larger variety of habitat types within the wood (different growth stages) and, therefore, the total number of bird species supported by the wood is likely to be greater than that generated by a selection system.
4 The complex foliage profile and high foliage volume of selection systems may lead to higher overall densities of birds and of species needing a dense shrub layer, for example Blackcap.

Group-felling occupies an intermediate position in this continuum of woodland patchiness (Fig. 6.4). Such a wood consists of a mosaic of small patches, typically some 50 m in diameter, at various stages of growth. This structure is likely to be more attractive to birds dependent on young-growth than selection woods, but the open areas are still small compared with those of clear-felling. Some birds of young-growth, such as Chiffchaff, will probably benefit from them but others, such as Tree Pipit, may not. In shelterwood, the size of patches is often similar to that of clear-felling. The early stages of shelterwood, however, support a very different bird community from that in young clear-fells. The retention of mature trees in the shelterwood for a decade or so, while the new crop grows beneath, leads to a far richer bird life than in clear-fells (Ferry & Frochot, 1970; Smith, 1988). In the regeneration stands, there can be a remarkable mixture of mature forest birds, such as woodpeckers and Nuthatches, and scrub birds including several species of warblers.

Selection systems are dependent upon skilful thinning which is a technique with wide potential for enhancing woodland bird communities, especially on nature reserves. In some circumstances, thinning can be used to stimulate regeneration so that a multilayered woodland structure is created. The aim would be to find an optimum balance between the density of mature trees and canopy openness. This type of careful manipulation could produce woods that are extremely rich in both mature forest birds and young-growth birds (Fuller, 1990). The effectiveness of promoting regeneration through selective thinning depends on the tree species involved. Light-demanding trees, such as oak, generally require far heavier thinning than shade-tolerant trees, for instance beech. In many woods, felling small groups of

Fig. 6.5. The Blackcap is a characteristic bird of any woodland which has a reasonably dense understorey. Territories in mature high forest are often associated with areas of dense undergrowth along rides or at the external edges of the wood. In coppice, the species typically colonises the regrowth at about the time of canopy-closure. Eric and David Hosking.

trees, perhaps coupled with thinning, may be a better approach than thinning alone. Some potential uses of thinning in derelict coppice are discussed below.

An outline of breeding bird communities in coppice

What general features distinguish the bird communities of actively managed coppice from those of broadleaved high forest? The rotations of high forest are much longer than those of coppice with the result that the proportion of the wood under young-growth is appreciably larger in coppice (Chapter 1). In clear-felling systems, there may be far less continuity of young-growth than in coppice where felling typically occurs at more frequent intervals. This means that coppice offers greater amounts of potentially suitable habitat for those breeding birds, notably several species of migrants, which generally prefer the early stages of forest succession. In addition, coppice seems to produce a habitat structure that is more consistently suitable for breeding migrants than that in young high forest. Within a few years of being cut, well managed coppice typically develops the thick bushy vegetation much favoured by several species of warblers and Nightingale. In high forest, the development of a comparable vegetation structure can be rather slower and is usually dependent on colonisation by non-crop trees such as birch. What lowland high forest may lack in summer visitors it often makes up for in hole-nesters. Unless densities of standard trees are very high, which is not good management practice (see below), densities of hole-nesters are generally rather sparse in coppice. Woodpeckers, Nuthatch and Blue Tit generally attain higher densities in high forest and wood-pasture, especially where oak is common, than in coppice.

The act of cutting coppice triggers a metamorphosis in the vegetation. The ground is initially bare, but assuming a reasonable density of stools, a profuse field layer develops in the second year. The density of foliage in the 2 m closest to the ground, peaks at 3 to 5 years after cutting, but this becomes shaded out soon after the canopy has closed, typically after 6 to 8 years. Until the next felling, the understorey remains sparse and, apart from a continued growth of individual trees, the structure is relatively stable. This highly predictable sequence of events is matched by large changes in the breeding bird communities, similar to those outlined for developing scrub, broadleaved high forest and conifer plantations. The following account is based mainly on information in Fuller (1992). It is only applicable to coppice in the lowlands; relatively little is known about responses of birds to the coppicing of sessile oak in the west and north, though it would be surprising if they were greatly different from those in the lowlands. Three broad phases in the bird communities of coppice can be recognised: the *establishment, canopy-closure* and *maturation* phases. The structure of the coppice in these phases, together with some of the bird species most strongly associated with them, are indicated in Fig. 6.6. The canopy-closure phase holds the greatest densities of warblers: this is the period when the canopy is slightly open or has very recently closed, yet the density of the field and shrub layers is still high. Once this low vegetation has been shaded out, the numbers of summer visitors decline rapidly, and in lowland English coppice the bird community often becomes dominated by Robins and tits. In old oak coppice in Scotland, Williamson (1976) reported that the Chaffinch was the most abundant breeding species.

This is a highly simplified view of coppice bird communities, for there is much variation from one wood to another in the timing and the strength of the responses shown by many species. Two major sources of this variation are the way the coppice is managed, which is the subject of the next section, and the tree species composition. Approximately half the area of surviving worked coppice is of sweet chestnut, though many traditional coppices in lowland England are more or less pure hazel, or mixtures of trees such as ash, hazel, field maple, sallow and birch. The bird communities of sweet chestnut coppice are substantially different from those of other types of coppice. In the establishment phase, Yellowhammers, Tree Pipits and Linnets are more frequently encountered in chestnut than in mixed coppice. It is possible that these birds are favoured by the drier soils on which

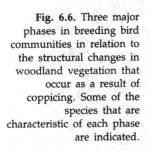

Fig. 6.6. Three major phases in breeding bird communities in relation to the structural changes in woodland vegetation that occur as a result of coppicing. Some of the species that are characteristic of each phase are indicated.

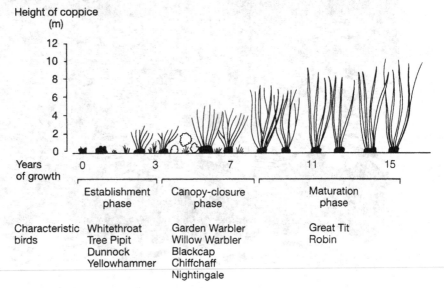

much chestnut is grown. Though Willow Warbler is a common bird in the canopy-closure phase of chestnut, other warblers appear to be less abundant than in mixed coppice. It is also unusual to find Nightingales breeding in pure chestnut coppice. Why should summer visitors be less numerous in chestnut than in mixed coppice? One possibility is that there may be less insect food available in chestnut. Another is that the structure of chestnut coppice may be less attractive to species that require a vigorous shrub layer, for chestnut is typically grown as a dense monoculture which rapidly shades out the low foliage.

The variable response of birds to coppice management is exemplified by the Nightingale. Coppice is widely considered to be a preferred habitat of the species in England (e.g. Bayes & Henderson, 1988). Yet its occurrence in coppiced woods is extremely patchy, even where underwood of a suitable age (say of 4 to 9 years of growth) is available. Not only does the bird eschew chestnut, but it is absent from a high proportion of woods with apparently suitable mixed coppice. Nonetheless, some actively coppiced woods in East Anglia and south-east England, do hold extremely high densities of Nightingales. It is something of a mystery as to why these particular sites are so favoured. It is possible that the reasons lie not so much with current habitat quality, but with the behaviour of the birds themselves. Much coppice has become unsuitable for the species through neglect. Even much apparently suitable coppice will have gone through prolonged breaks in management causing the birds to become extinct at these sites. If individual birds show strong fidelity to particular sites, or are attracted to sites already holding birds (Fig. 4.10), then the species may have difficulty in recolonising former sites, even when suitable habitat is restored. Some long-distance migrants, for example Willow Warbler, do have the capacity to colonise new habitats very rapidly, but it cannot be assumed that the same holds true for the Nightingale. This is merely a hypothesis; time will tell whether re-establishing coppicing on nature reserves is beneficial to the Nightingale.

The influence of coppice management on birds

There are many ways to coppice a wood. Coppice systems differ in their rotation lengths, densities of stools, the size of individual coppice compartments, and densities of standard trees. To some extent these parameters depend on the tree species. For example, ash and oak are grown on much longer rotations and wider stool spacings than hazel, while chestnut is grown more often without standards than other types of coppice. Here I summarise the influence of two of the most

Fig. 6.7. A Tree Pipit singing at the edge of a freshly cut panel of sweet chestnut coppice. Chris Rose.

important aspects of coppice management for breeding birds – rotation length and standards. For more detail and management recommendations see Fuller (1992), Fuller & Warren (1993), Fuller & Peterken (1994).

The rotation length determines the proportion of the wood that is covered by young- and middle-growth coppice. In general, the shorter the rotation, the greater will be the amount of potentially suitable habitat for summer visitors. Some areas of old coppice are desirable because these may be especially beneficial to tits; for instance they might harbour the only pairs of Coal, Marsh or Willow Tits in an actively coppiced wood. Very long rotations, however, do not generally result in more diverse woodland bird communities, because within coppice there are few birds that are absolutely restricted to the oldest stages. Indeed, mature coppice where the canopy has been closed for several years, is generally very poor in terms of densities of birds and numbers of species. Where a rich bird community is sought within a coppiced wood, it is best to adopt rotations that will create a preponderance of underwood in the early and middle stages of growth, but also to leave some patches to grow on well beyond canopy-closure.

The effects of standard trees on the bird life of coppice are especially important. The addition of an extra stratum of vegetation, in the form of mature trees scattered among the coppice stools, has a generally beneficial effect (Fig. 6.8). Standards provide nest sites for hole- and canopy-nesting birds where they would otherwise be lacking. The tree foliage also offers feeding sites for Chaffinches, tits and other insectivorous birds. However, standards also have an indirect effect on birds by casting shade, which can suppress the growth of the coppice itself. This becomes a problem where the density of standards is so great that the coppice is unable to develop the vigorous, bushy structure that many summer visitors require. In most coppiced woods, there is likely to be a particular structure where there are sufficient large trees to provide habitats for hole-nesting and canopy-feeding birds, but not so many as to make the coppice unsuitable for warblers and Nightingales. Attaining this ideal balance between mature trees and dense coppice growth is difficult to achieve and it is best to err on the side of leaving too few, rather than too many standards. As a general rule, a coppiced wood should have no more than ten large standards per hectare. There is no point in trying to grow coppice under a high density of standards. To do so compromises both the market quality of the underwood and the special conservation interest of coppiced woodland – the species associated with unshaded, early successional stages. If one wants to create a woodland habitat dominated by large trees, a high forest approach should be adopted.

Derelict coppice

Large areas of formerly coppiced woodland lie abandoned, including much hazel and mixed coppice in southern and central England, and sessile oak coppice in the west and north. Derelict coppice long retains its characteristic multi-stemmed form, though much of it will gradually assume a high forest structure through self thinning. From a conser-

Fig. 6.8. Hazel coppice growing beneath oak standards, Hornhill Wood, Worcestershire. The hazel stool in the foreground was probably cut some five years previously. Such vigorous regrowth can be attained only where the density of standards is not too great and deer are not a problem. Note the absence of coppice growth in the immediate vicinity of the large tree. David M. Green.

vation viewpoint there is little reason to retain these woods in their present structure. They have long since lost the bird communities and other wildlife associated with young, actively managed coppice. As described above, overmature coppice is a poor habitat for birds. The options for their future management are essentially either to recoppice, or to convert to high forest. The long-term implications of these two routes for the bird communities will be evident from the earlier parts of this chapter where comparisons were drawn between high forest and coppice.

Restoring a coppice regime is not always the best approach. Labour costs are high and the markets are small and specialised. Furthermore, much derelict coppice is unlikely to regenerate well because of low densities of viable stools and increasingly severe pressure from deer (Fuller & Peterken, 1994). In many cases, gradual conversion of the woodland to a high forest structure is more desirable. This would lead to bird communities that are more interesting than those of both derelict coppice and of restored coppice that did not regenerate

Fig. 6.9. Derelict hazel coppice, with scattered oak standards, Hill Wood, Worcestershire. This coppice has been allowed to grow on well beyond its normal felling age. The open structure is typical of coppice once it has closed canopy; in the absence of further management the structure of the wood will remain much the same for many years. The densities of birds breeding in old coppice are low and typically the most abundant species are Robin and several species of tits. David M. Green.

vigorously. The way chosen to achieve a conversion to high forest depends on the type of woodland. An appropriate route in pure hazel, for example, would be to replant gradually through a programme of group-felling. By contrast, promoting high forest by thinning the coppice to remove all but one or two stems on each coppice stool (this process is called singling) is a realistic proposition for lime and some mixed coppice. It would, however, be risky to undertake extensive singling in exposed western oakwoods, on steep slopes, where there was a real danger of serious windthrow. Singling can rapidly alter the bird community of derelict coppice because the opening of the canopy can stimulate growth of the field and shrub layers (Fuller, 1990). Species such as Wrens and Blackcaps can increase their numbers over quite a short period in response to the thinning, though singled coppice carries lower densities of breeding warblers than well-managed active coppice. Another approach is to leave the task to nature. The woodland will gradually assume a high forest structure, though this may take a long time. The woodland could be eventually managed or left to develop some of the characteristics of natural woodland. There are many areas of derelict coppice, especially within nature reserves, which would be suitable sites for developing substantial tracts of near-natural woodland. Woods with many old trees are very rare habitats indeed in Britain. In the long-term, derelict coppice could form the basis of a substantially increased area of old-growth woodland.

7

Upland woods and wood-pasture

Man has allowed his animals to graze in European woods, or used the products of woods as animal fodder, for thousands of years. These traditions have all but disappeared in the lowlands, though the legacy of a long history of grazing can still be seen in the vegetation of some woods such as parts of Epping Forest, Windsor Forest and Burnham Beeches where large numbers of old pollards survive. Very few woods continue to be actively grazed by domestic animals, the main examples are in the New Forest. Despite the decline of wood-pasture, grazing pressure is perhaps greater now than at any other time in the lowlands. This paradox arises because numbers of deer are increasing so rapidly that they are transforming the structure of many lowland woods. Deer are now a serious obstacle for the successful regeneration of trees within both high forest and coppice systems.

The great majority of upland woods are open to grazing by deer and sheep and there has long been concern about the widespread lack of regeneration. In winter, many of these woods are now intentionally used for feeding and sheltering sheep. The structure of these woods is not dissimilar to that of some lowland wood-pastures, though the trees rarely reach such impressive dimensions and pollarding is seldom, if ever, practised. Wood-pasture, as a system of managing woodland, is described in Chapter 1.

This chapter describes the bird life of heavily grazed broadleaved woodland and reviews the effects of grazing of woodland on birds. The problem of overgrazing in native pinewoods is covered in Chapter 8. It is difficult to exaggerate the importance of grazing as a force in shaping both the structure and species composition of woodland vegetation. By removing a substantial part of the woodland foliage, grazing has profound implications for birds and other woodland animals. Unless stated otherwise, I use the term 'grazing' here in a general sense to embrace all the actions of large herbivores, including the browsing of tree and shrub foliage.

Upland birchwoods

Woods of more or less pure birch are the commonest type of woodland in north-west Scotland. In the absence of grazing, birch can quickly establish itself because the seeds are widely dispersed by the wind. Birchwoods can regenerate themselves if grazing is light or intermittent. If severe grazing pressure is sustained, however, existing birchwoods can quickly thin out and disappear because the trees are not generally long-lived. Alternatively, the boundaries of individual woods may gradually shift in response to local relaxation or intensification of grazing. Heavily grazed birchwoods are devoid of any foliage in the 2 m or 3 m closest to the ground. Light grazing, however, can allow shrubs to develop beneath the birch canopy, most commonly rowan, hazel, birch and juniper. Highland birchwoods with contrasting structures are shown in Figs. 4.7 and 4.8.

By far the two commonest breeding birds in Scottish birchwoods are Chaffinch and Willow Warbler, which together may account for as much as half the total density of birds (Yapp, 1974; Bibby *et al.*, 1989). In this respect, the birds of the Scottish birchwoods are similar to those of the great birch forests of northern Fennoscandia where Willow Warbler and Brambling (ecologically, the replacement of Chaffinch in these northern forests) typically contribute half of the total birds, with Willow Warbler by far the commoner species (Enemar *et al.*, 1965). Other common breeders in Scottish birch include Wren, Robin, Blue Tit, Coal Tit, Great Tit, Tree Pipit, Redstart, Spotted Flycatcher and Wood Warbler. Pied Flycatcher is a surprising absentee from most birchwoods. In Scandinavia, Pied Flycatchers commonly nest in birch forest.

The Scottish birchwoods carry a very distinctive bird community, but it is rather poor in species. Furthermore, a small number of species contribute a high proportion of the total birds. Yapp (1974) considered that the diversity of birds in Highland birchwoods was less than half that found in most English broadleaved woods. His measure of diversity was based on the probability of two individuals being of different species, when drawn at random from all those counted. Species that are common in many lowland woods but relatively scarce in upland birch, even where these woods apparently lie within their distributional ranges, are Woodpigeon, Blackbird, Song Thrush, Chiffchaff, Blackcap and Garden Warbler.

Birchwoods are far less common in the uplands south of the Highlands, but where they do exist the bird communities are similar to those described above (Yapp, 1962). Chaffinch and Willow Warbler remain by far the most abundant species, though the latter may be less numerous in English and Welsh woods than in northern Scottish woods. In a small area (5 ha) of mature birch in west Wales, Hope Jones (1972) found that the three most abundant breeding birds were Willow Warbler, Chaffinch and Wren.

Despite their relative simplicity, and constancy of species composition, the breeding bird communities of Highland birchwoods are not entirely uniform (Bibby *et al.*, 1989*b*). Woods in the western Highlands generally support more impoverished bird communities than those in the east, both in terms of numbers of species and overall densities of birds. Several species, including both Willow Warbler and Chaffinch, are commoner in eastern birchwoods but Wren is an exception in showing the opposite pattern. These east–west trends cannot be readily explained by variation in habitat structure. It seems likely that they reflect underlying gradients in land productivity and climate, which in turn, influence the amount of insect food available for birds. The density of the shrub layer is another important factor for the birds of birchwoods. Numbers of some species, such as Willow Warbler and Wren, are higher where the shrub layer is better developed (Fig. 4.8), but Tree Pipit and Redstart prefer more open areas (Fig. 4.7). Woods that have a mixture of shrubby and open areas tend to offer suitable habitat for more breeding species than the more uniform woods.

There is little published information on the winter bird populations of Scottish birchwoods. However, once the summer visitors have departed, the bird life in these woods must be extremely sparse. Most of the sites are very exposed and it is likely that many 'resident' individuals move out of the woods to spend the winter in lower altitude woods or in other habitats.

Upland oakwoods

Oakwoods have a far wider distribution in northern and western Britain than do those of birch. Outside the north-west Highlands, where birchwoods are commoner, oakwoods form the main semi-natural woodland within the upland regions of Britain, in which I include the south-west peninsula of England. Major concentrations of upland oak occur in western Scotland, the southern Highlands, the Lake District, Wales, Devon and Somerset. The oakwoods in the Wyre Forest and the Forest of Dean lie on the fringes of this upland region. The upland oakwoods are mostly composed of sessile oak whereas lowland oakwoods usually, but not exclusively, consist of pedunculate oak. Much upland oak has a history of having been coppiced, often retaining the characteristic multi-stemmed growth form. As with birch, the majority of upland oakwoods have survived on valley slopes and steep hillsides. While oaks dominate these woods, they may contain several other tree species including birch, ash, alder, rowan and hazel. Just as for birchwoods, the structure of the woodland is strongly affected by grazing. Most upland oakwoods show a clear browse line, with extremely sparse field and shrub layers.

The following account outlines the main characteristics of the bird life in upland oakwoods. For more information see Yapp (1962), Simms (1971), Edington & Edington (1972), Hope Jones (1972), Williamson (1972a; 1974), Gibbs & Wiggington (1973), and Stowe (1987).

In general, the numbers of species and individual birds breeding in upland oak are lower than those in a comparable area of lowland oak. A major exception is the New Forest where the bird life of the oakwoods is similar in several respects to that of upland woods (see below). The assemblages of migrants are fundamentally different in upland and lowland oak. Upland oakwoods are typified by large populations of Pied Flycatchers, Redstarts and Wood Warblers, which are scarce in lowland pedunculate oak. Tree Pipits are also more abundant in upland than lowland broadleaved woods. Residents, on the other hand, show no striking differences between the uplands

Fig. 7.1. Sheep-grazed sessile oak woodland, Borrowdale Woods, Cumbria. Peter Wakely, English Nature.

and lowlands. Sessile oakwoods growing on Welsh hillsides support all the resident woodland bird species that one would expect to find 50 or 100 miles to the east in the English Midlands.

The bird communities of upland oakwoods are considerably richer than those of upland birchwoods. In those areas of Scotland where both types of wood are present, oak holds more bird species (French *et al.*, 1986). Furthermore, Chaffinch and Willow Warbler do not dominate in oak to the same extent as in birch. Four or five species stand out as numerically dominant in Scottish oakwoods: Chaffinch, Wren, Robin, Willow Warbler and, perhaps, Blue Tit. Three or four of these species typically contribute approximately half of the total breeding birds. The bird communities of the Welsh and English upland oakwoods are richer still, and more complex, than those of the Scottish. All the species characteristic of the Scottish woods are abundant further south, but they are supplemented by Pied Flycatcher, while Redstart and Tree Pipit are more frequently among the most numerous breeders.

The bird life of upland oakwoods not only varies between regions, but it can also vary substantially from one wood to another within the same region. Some of this variation is associated with differences of habitat structure, as shown by Bibby & Robins (1985) for woods in south-west England. The complexity of the vegetation structure, both the canopy patchiness and the vertical foliage profile, was linked to variation in the composition of the bird communities in this sample of woods.

The upland woods are generally inhospitable habitats for birds in winter. Yapp (1959; 1962) made transect counts of birds wintering in sessile oakwoods in England and Wales. The most northern woods, those in Cumbria, were almost devoid of birds but he reported that winter densities in Welsh and south-western oakwoods were not greatly below those found in summer. Nonetheless, in all woods there were substantial changes in species composition. Chaffinches and Yellowhammers were absent from all woods in winter and Robins were also much reduced. Somewhat contrasting conclusions were reached by Hope Jones (1975) who made a detailed study of a sessile oakwood in western Wales over four winters. Birds were counted at weekly intervals using a transect technique. The main findings were that: (a) two-thirds of the birds counted were tits with Blue Tit contributing a third of all birds; (b) numbers of birds declined as the winter progressed; (c) numbers of birds were lower on the upper slopes; (d) the overall density of birds in winter was low – approximately 10% of the breeding density on the upper slopes and 30% on the lower slopes; (e) for most birds the wood formed just a part of their winter range.

Lowland wood-pasture

Wood-pasture may be obsolete but relics of these systems survive in the lowland landscape (Harding & Rose, 1986). The remnants of lowland pasture woodlands take several forms but they have one thing in common – they are typified by the presence of large, old trees, usually oaks or beeches (Fig. 1.9). Many of these trees have been pollarded at some stage in their lives and they are often hollow. Compared with managed high forest, pasture woods are typified by large amounts of dead wood (Smith *et al.*, 1992). This wood exists in

Fig. 7.2. Throughout England and Wales, the Pied Flycatcher is a characteristic bird of grazed upland oakwoods. North of the central lowlands of Scotland, however, there is much similar habitat where the species is absent. Chris Rose.

various stages of decay, as standing dead trees, dead boughs and heartwood on living trees, and as fallen wood. The ancient trees often provide unique habitats for certain fungi, epiphytic lichens and bryophytes, and for many invertebrates dependent on dead or dying wood. The trees act as refuges for many of these species, which were presumably far more widespread in the primeval forest, but whose survival has been possible in relic populations where suitable conditions have persisted.

There are close parallels between the bird communities of lowland wood-pasture and upland grazed woods, presumably because their habitat structures are so similar. Both typically lack a vigorous undergrowth, and birds associated with a dense shrub or field layer are generally scarce or absent. Few warblers are present, with the exception of Wood Warbler. Ancient wood-pastures are, however, generally rich in birds that feed on invertebrates associated with dead wood. The densities of hole-nesting species are high in much pasture woodland but they are not necessarily greater than those in mature high forest. This is the case in the New Forest where Smith *et al.* (1992) counted birds in both types of woodland. The plantations were of oak ranging in age from 70 to 185 years. Blue Tit and Chaffinch were by far the two most numerous species in both types of woodland. Only two species, Robin and Wood Warbler, occurred at higher densities in the pasture woods and the two habitats carried very similar overall densities of birds. Furthermore, there was no indication that the amount of dead wood was related to the numbers of any species, with the possible exception of Redstart.

In an earlier comparison of a wood-pasture stand and a mature oak plantation in the New Forest, Irvine (1977) found that hole-nesters made up a similar proportion of the breeding community in both habitats but their density was somewhat lower in the plantation. It is impossible to draw conclusions about general differences between wood-pasture and plantations from this study because the number of plots was too small. In any case, counts of birds may not reveal the full story. It would be interesting to know whether there are differences in the breeding success of Nuthatch, Treecreeper and woodpeckers between these two habitats.

Not all surviving wood-pasture has more or less continuous tree cover, as in many of the unenclosed stands of the New Forest. In some cases overgrazing, coupled with overexploitation of the trees themselves, has been so severe that remaining trees are widely scattered, sometimes interspersed with scrub and younger trees. Some woods may even have been transformed into open commonland or heath. Other wood-pastures have been converted into managed high forest, though some ancient trees may still endure. Others have been incorporated into landscaped parks and a handful survive in the form of medieval deer parks, many of which were originally a mixture of wooded and open land. Windsor Forest and Great Park contain a variety of these wood-pasture habitats, all with spectacularly gnarled old oaks, often termed *hulks*. Counts of breeding birds at nine sites in three different habitats at Windsor showed that open parkland, with widely spaced ancient trees scattered across grassland, carried extremely low numbers of species and individual birds (Fig. 7.3, sites 1–3). The commonest breeders in these parklands were Stock Doves and Jackdaws, with just a smattering of small songbirds such as Blue Tits. The density of hulks in the parkland was similar to that in the semi-natural, open-canopy pasture woodland where there were patches of scrub (Fig. 7.3, sites 4 and 5). Yet the numbers of birds in the parkland were substantially lower than those in the open-canopy wood-pasture, which carried similar numbers to the closed-canopy broadleaved woods (Fig. 7.3, sites 6–9). This suggests that the birds were responding to the overall structure of the habitat and not just to the number of hulks.

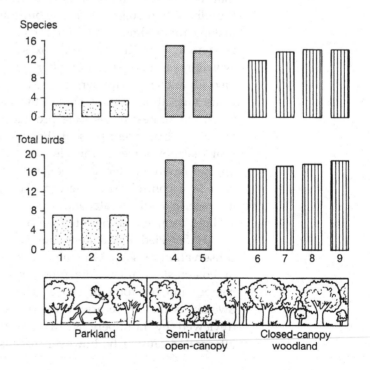

Fig. 7.3. Numbers of species and individual birds in three habitats in Windsor Forest and Great Park. Birds were counted in the spring at nine sites, each containing 6–10 point count stations. Means were calculated for each site and data have been combined from two years (1987–8). See text for more details of the habitats. The counts were made by E.E. Green.

Migrant songbirds in grazed woodland

Grazed woodland is the main habitat for Pied Flycatchers, Redstarts and Wood Warblers in Britain. The distinctive character of the bird life in western sessile oakwoods derives largely from these summer visitors, together with Tree Pipit. Each of these species prefers woods that are rather open beneath the canopy. This partly explains their predominance in western and northern Britain where a high proportion of woods are of suitable structure as a consequence of heavy grazing. A completely different pattern of habitat selection is shown by the common migrants in lowland woods (Nightingale, Garden Warbler, Blackcap, Willow Warbler, Chiffchaff) whose numbers are generally highest where there is a dense shrub layer or scrub. Where open woods exist, Redstarts, Wood Warblers and Tree Pipits can be found in the lowlands. However, the Pied Flycatcher is virtually absent from the lowlands, even from apparently suitable woods such as the New Forest wood-pastures. It is possible that this bird is limited by climate, for its English and Welsh distribution closely matches that of high rainfall areas (Yapp, 1962), though this relationship does not hold in Scotland.

The Pied Flycatcher has been the subject of intensive research, partly because it so strongly prefers nest boxes to natural tree-holes that the entire population of a wood can be enticed into boxes (Lundberg & Alatalo, 1992). The birds are also easily and safely caught, making them almost ideal for carrying out experiments on ecological behaviour. Introducing nest boxes to a wood leads not only to a switching of nest site, but often to a large increase in numbers of Pied Flycatchers. Numbers can rise way beyond any natural densities in no time at all. This is shown by work in Swedish sub-alpine birch forest where nest boxes were provided following 2 years of censusing when no boxes were present (Enemar & Sjöstrand, 1972). In the first year that boxes were available, the Pied Flycatcher density rose from approximately 0.1 territory per hectare to approximately 2.5 per hectare. In a control area where no boxes were provided there was no increase in Pied Flycatchers.

Huge numbers of nest boxes have been erected in western British woods for Pied Flycatchers, not so much for research but for bird conservation reasons. One might ask whether the consequent increase in hole-nesters affects the other bird species in the wood through competition for food or space. The evidence from three studies in Sweden is that there are no effects on the numbers of other species (reviewed by Lundberg & Alatalo, 1992). These results should be interpreted cautiously, however, because the breeding success of birds was not examined. Furthermore, it cannot be assumed that the situation is the same in Scandinavian woods as in upland British woods. It would be particularly interesting to know more about potential interaction between Pied Flycatcher and Redstart, which in Welsh oakwoods appear to occupy very similar foraging niches (Stowe, 1987). These two species feed at all heights from ground to canopy, whereas Wood Warbler feeds mostly in the canopy and Tree Pipit almost exclusively on the ground.

The overall density of migrants is often close to that of residents in upland oakwoods. This ratio of migrants to residents is higher than in other British woodland bird communities. This arises both because absolute densities of migrants are high, and because densities of residents are relatively low in upland woods (Fuller & Crick, 1992). The relatively large populations of migrant birds in upland woods may be a consequence of competition, both direct and indirect, with resident birds (see Fuller & Crick). The winter populations of resident birds in upland woods are low, perhaps leading to much lower predation pressure on invertebrates, compared with lowland woods. This could result in relatively large food resources being available in the spring for migrants to exploit, particularly because numbers of breeding residents are low. Upland woods with their sparse shrub layers and low variety of tree species are probably poor breeding habitats for many residents. It has been suggested that migrants are competitively inferior to residents and that this may influence their habitat use (O'Connor, 1985). If this is true for woodland songbirds, then the low numbers of residents may allow more 'ecological space' for migrants in upland woods. There have been no convincing studies of competition between migrants and residents in British woods, but Swedish research on Collared Flycatchers demonstrates that such effects are a real possibility (Gustafsson, 1987). An experiment was conducted in which the density of breeding tits was greatly reduced in half of the nest box study plots but not in the remainder. In the areas with reduced tit numbers, Collared Flycatchers bred more successfully and contributed more birds to the breeding population in the subsequent year. The control areas did not show these effects. It is likely, therefore, that there was competition for food between the tits and flycatchers.

Interestingly, British upland woods are analogous to central European woods, which also carry large populations of Wood Warblers and flycatchers. The European woods have few overwintering residents due to their inclement climate, and also rather low densities of breeding tits.

Effects of grazing of woodland on birds

The effects that grazing has on bird populations in British woodland have hardly been examined, but we know from North American work that there can be very large impacts on woodland birds (Dambach, 1944; Casey & Hein, 1983). Two comparisons of breeding birds in grazed and ungrazed Scottish oakwoods (Williamson, 1972a; 1974) did not reveal any striking differences in the overall density or richness of the bird community. The relative abundances of some species appeared to differ between grazed and ungrazed plots, but not in a consistent way between the two studies. Far more comparisons of this type are needed, with measurements of vegetation as well as counts of birds, before any general conclusions can be drawn about the effects of grazing in upland woods on birds. However, counts of birds in two winters in the Forest of Dean indicated that ungrazed oak carried larger numbers of species and individual birds than grazed oak (Hill *et al.*, 1991). Furthermore, in Wytham Wood near Oxford, increasing numbers of fallow and muntjac deer have greatly reduced the field and shrub layers in recent years. This habitat change is thought to have contributed to the loss of the Nightingale from the wood and has probably severely affected other bird populations (Gosler, 1990).

The published information may be scant but there is no doubt that the effects of grazing of woodland on birds are complex and far-reaching. Grazing can potentially affect birds in three ways: by altering the structure of the vegetation, by altering the plant species composition and by altering the amount and types of food available. I shall take each of these potential effects in turn, though they are closely inter-related. This must be regarded as the most general of accounts because the complexities of grazing are enormous and significant variations will arise according to the type of woodland, the types of animals involved, the density of animals, and the constancy of grazing. Clearly, grazing is not a simple 'all or nothing' process. For more information on grazing in woods see Putman (1986), Tubbs (1986), Mitchell & Kirby (1990). Possible implications of different levels of grazing are summarised in Fig. 7.4.

1 *Woodland structure* The most obvious impact of heavy grazing is the virtual elimination of the field and shrub layers below a sharply defined browse line at 1.5 m to 2 m (Pigott, 1983; Putman *et al.*, 1989; Mitchell & Kirby, 1990; see Fig. 7.5). Given that many species in British woods depend on the shrub layer for feeding and nest sites, one would predict that heavy grazing would cause a reduction in the overall density and richness of bird communities. Several

examples have been given in earlier chapters of how complexity of the foliage profile is generally associated with richer bird communities. More important perhaps, are the likely changes in bird species composition, for some birds potentially benefit from grazing while others do not. The migrant birds provide obvious examples with Pied Flycatcher, Redstart, Wood Warbler and Tree Pipit apparently preferring heavily grazed woods, but Nightingale, Blackcap and Garden Warbler requiring the type of vegetation structure that does not generally occur in heavily grazed woods.

A change in vegetation structure could have quite subtle effects on birds. Ground-nesting birds, and those using bramble, may be more susceptible to predation in heavily than moderately grazed woods because their nests are easier to find where the field layer is extremely sparse. Nests are also more likely to be trampled where the density of herbivores is high. It is conceivable that variations in grazing intensity may cause some woods to be population sinks (where the population is maintained by immigration from other areas) while other woods may act as population sources (where a surplus of young birds is produced).

2 *Plant composition* Large herbivores have major impacts on the relative abundance of different plant species but these vary from one animal to another. Sheep, for example, are more selective grazers than cattle. Some animals, for example red deer, are more inclined to browse trees and shrubs than say sheep or cattle. The exact impact of any animal depends to a large extent on which plant species are available to it. An animal may concentrate its attention on a preferred plant when it is available, but in the absence of that plant

Fig. 7.4. Some possible implications for birds due to different intensities of grazing pressure in woodland.

Grazing pressure	Effects on vegetation	Possible implications for birds
High	No shrub layer Sparse field layer No tree regeneration Bare soil patches Fewer shrub species	Few shrub-nesting birds Heavy predation on ground-nesters Reduced food resources (berries, foliage invertebrates, small mammals)
Moderate or Low	Patchy shrub layer Diverse field layer Few bare soil patches	Nest sites and feeding sites available for a wide range of species
None	Dense shrub layer Diverse field layer	Many shrub-nesting birds Few open woodland birds Few ground nesters

it may switch to another, with entirely different consequences for the resulting vegetation. Despite this complexity, it appears that heavy grazing can lead to a reduced diversity of plants in both the field and shrub layers with the loss of several plant species. This is the case in the unenclosed New Forest woodland, where there has been almost total loss of blackthorn, hawthorn, hazel, field maple, willows and bramble (Putman *et al.*, 1989). It would be very surprising if such severe impoverishment of tree and shrub species had not affected the bird communities in ways additional to the changes in woodland structure. One would predict that bird communities should become less varied with the loss of tree species (Chapter 4). The variety of foods available to birds will be greatly reduced with loss of plant species.

3 *Food availability* The loss of a substantial proportion of the low foliage and the selective browsing of certain shrubs has major indirect implications for birds. Presumably the abundance of many insect groups associated with low woodland foliage will be greatly reduced, with implications for insectivorous songbirds. There are likely also to be fewer berries produced on low shrubs. In the New Forest,

Fig. 7.5. The impact of deer grazing on woodland vegetation. Deer have been excluded from the area to the left of the deer-proof fence which can be seen in the centre of the photograph. Within the ungrazed area there is vigorous growth of bramble, but there is no field layer in the grazed area. Roudsea Wood National Nature Reserve, Lancashire. Peter Wakely, English Nature.

formerly grazed areas which have been fenced to exclude deer and domestic stock have much higher populations of small mammals than the grazed woodland (Putman *et al.*, 1989). This suggests that there may be lower amounts of food available for predatory birds, such as owls and Buzzards, in grazed than ungrazed woods. Buzzards in the New Forest breed more successfully in years following high production of tree and shrub seeds (Tubbs & Tubbs, 1985). A likely explanation for this pattern is that the seedfall leads to an increase in rodents which otherwise remain at a low level because the severe grazing pressure reduces the cover and amount of food available to them.

Conservation issues and management of grazed woodland

The main conservation issues in upland woods and wood-pastures are about conservation of fungi, epiphytes and dead wood insects, rather than birds. Nonetheless, the bird communities of these woodland types are highly distinctive and any management proposals should consider the likely effects on birds. Lowland wood-pastures are so rare that they are among the most important conservation sites we have and it is vital that these places and their ancient trees are protected.

There are two main management needs (Fuller & Peterken, 1994). The first is relevant to those sites supporting old trees with associated scarce fungi, epiphytes and saproxylic insects. The survival of these populations is utterly dependent upon the continuity of dead wood. It is essential, therefore, that steps are taken to create the next generation of old trees. This management action is not of particular significance for bird populations, though it could be that some species such as Redstart are mainly associated with ancient trees in some wood-pastures (Fig. 7.6). The second management need is, however, of great importance in determining the nature of the bird community. This concerns how best to encourage regeneration without compromising the special conservation interest of the habitat. This question is relevant to both wood-pasture and upland woods.

Woods are more resilient to being grazed than one might imagine. It only needs very occasional periods of successful regeneration for a wood to maintain itself. Despite a long history of high grazing pressure, the New Forest woods have undergone intermittent phases of reduced grazing which have been sufficient to permit regeneration (Tubbs, 1986). The problem may be most serious in the case of birchwoods where the trees are much shorter lived than oak or beech. Whether or not grazing poses a real threat to woodland, there is no denying that many woods are overgrazed to the extent that they have an extremely impoverished fauna and flora. Relaxation of grazing would, in many woods, lead to much richer biological communities but there is much to learn about how this might be best achieved (Mitchell & Kirby, 1990).

Total cessation of grazing is unlikely to be satisfactory because this may result in rather uniform dense regeneration which would probably lead to loss of lichens and bryophytes. A dense shrub layer would

Fig. 7.6. A Redstart at the entrance to its nest-hole in a beech pollard. Chris Rose.

also reduce the quality of the habitat for Pied Flycatchers, Redstarts, Wood Warblers and Tree Pipits. It may prove possible to use low or moderate levels of stocking, or seasonal, or episodic grazing to create a woodland consisting of a mosaic of heavily grazed patches and regeneration patches. Such a varied habitat structure may support richer bird communities than either heavily grazed or ungrazed woodland, because different patches of vegetation would suit different species (Fig. 7.4). It should also be remembered that trees can fail to regenerate for reasons other than grazing by large herbivores so that manipulating grazing may be only a part of the answer. In the case of oak, these additional factors can include defoliation of seedlings by caterpillars, consumption of seeds by small mammals, competition from bracken and grasses, and grazing by rabbits (Humphrey, 1992).

Even where the problem is overgrazing, it may not be necessary to introduce a new grazing regime. It could be that some thinning of the canopy, or felling small blocks of trees, would allow sufficient light to stimulate regeneration. This would not be an appropriate approach in lowland wood-pasture but it could have wide application in upland oakwoods, though care would need to be taken at exposed sites where there could be a risk of windthrow. An interesting example was given by Stowe (1987) who described how the density of oak saplings in a grazed sessile oakwood with a broken canopy was 3.5 times the density that was present in a wood that had been ungrazed for 25 years. Both these woods had far higher densities of oak saplings (by factors of 669 and 191 respectively) than did a sample grazed wood. The bird life of open-canopy woods with patches of dense regeneration could be extremely interesting. In his study of the birds associated with different stages of growth in sessile oakwoods, Hope Jones (1972) found that the highest densities of birds were in mature oak with gaps and scrub. The bird community in this habitat was a rich mixture of resident and migrant birds. It was particularly remarkable in holding good populations of the four classical sessile oakwood summer visitors – Pied Flycatcher, Redstart, Wood Warbler and Tree Pipit – alongside high numbers of scrub warblers. This suggests that, as far as migrant birds are concerned, it can be possible to have the best of both worlds through imaginative habitat management.

8

Coniferous woodland

Planting of coniferous forests has been one of the major changes in British land-use this century. The consequences for birds have been various and complex. Though detailed study is essential to understand the ecological effects of forestry, this is not just a scientific matter because it involves value judgements about the relative worth of different types of bird communities and species.

The main conservation issue has been the impact of forestry on birds of adjacent open moors and the loss of moorland through the planting of trees (Thompson *et al.*, 1988; Avery & Leslie, 1990). Strictly speaking, this subject lies outside the scope of this book but a few observations at this point are apposite. It is tragic that so much really special lowland heath and upland habitat has disappeared beneath conifers. This is not to imply that all the planting has been ill-placed. Many upland areas, now covered by conifers, had been severely impoverished by a long history of overgrazing and burning; generally these would have held low numbers of birds before planting. Nevertheless, in many cases, the environmental impacts could have been less severe if the plantations had been more carefully located (Avery & Leslie, 1990). While it is unlikely that national populations of any moorland birds are under threat of extinction from afforestation in Britain, it is undeniable that there have been substantial regional losses. The most visible example must be the Flow Country of Caithness and Sutherland where forestry has been imposed on a unique peatland landscape that is home to a remarkable collection of rare birds. The history and impacts of this massive change in land-use were dissected from rather differing standpoints by Avery & Leslie (1990) and Ratcliffe (1990). This is a clear case where the new bird communities colonising the plantations cannot, in any way, be regarded as compensation for the loss of a special upland community.

Afforestation eventually leads to characteristic moorland species such as Curlew, Golden Plover and Red Grouse being replaced by rather widespread woodland songbirds such as Goldcrest and Coal

Tit. This is a simplistic view, however, because the new forests do provide habitats for several interesting and scarce birds, for example Goshawk and Black Grouse. Furthermore, some moorland species may find a future habitat in forestry restocks, though so far the signs are not promising, particularly for waders. Coniferous forest will be a lasting feature in many parts of Britain and it represents a vast habitat for woodland birds. It makes much sense, therefore, to ensure that we capitalise on the opportunities it offers. For this reason the last part of this chapter is devoted to ways of improving plantations for birds.

The effect of planting conifers in existing woodland is an entirely different issue from that of afforestation. Here the comparisons are between bird communities in conifers and those of the woodland, usually broadleaves, which they replaced. The effect will depend on the tree species involved. For example, replacing oak with Norway spruce may lead to a greater reduction in bird numbers than replacing beech with Norway spruce. In the latter case, the conifers may even lead to an increase in numbers of birds, though the species composition would be different. Because bird communities undergo such large changes during the growth of plantations, it is only sensible to compare the bird communities of broadleaved and coniferous stands of similar ages. There is little point in comparing a 90 year-old stand of oak with freshly planted spruce. Many woods have been only partly planted with conifers, and in such cases the effect on bird communities may often be to increase the overall variety of birds using the wood.

The great majority of coniferous woods available to birds in Britain are plantations but there are two exceptions: the native pinewoods of Scotland and yew woods. The third native conifer, juniper, locally forms an open scrub and is the main constituent of the shrub layer in some northern pinewoods. Before considering the man-made conifer forests we will examine the bird communities in these two native woods.

Native pinewood

Only scattered fragments remain of the natural Scots pine forest that once covered much of the Scottish Highlands, but these harbour one of the most distinctive woodland bird communities. Three bird species have especially important populations in these forests: Capercaillie, Crested Tit and Scottish Crossbill, the latter being Britain's only endemic bird. In bird conservation terms these are the key species, but a wide range of other birds use the woods.

Studies at Abernethy Forest, Strathspey, show that the two commonest species are Chaffinch and Goldcrest but it is by no means
clear which are the next most abundant species. Hill *et al.* (1990) list
Spotted Flycatcher, Coal Tit, Crested Tit and Siskin, but Newton &
Moss (1977) found Wren, Coal Tit, Robin and Willow Warbler to be
among the six commonest species. These differences may have arisen
because different methods were used in these two studies. The Pied
Flycatcher has bred only sporadically in Scottish pinewoods, which
is surprising because in Scandinavia it commonly breeds in coniferous
forest, though not at such high density as in broadleaved woods.
Small numbers of Wrynecks have colonised a few areas of mixed
open pine and birch in the Highlands.

In general, the native pinewoods are richer habitats for birds than
mature pine plantations in the same region, in terms of their overall
bird densities and numbers of species. There is, however, much
variation in the structure of the native pinewoods which influences
the bird communities found in them. Three features seem especially
important in this respect. Some of the woods are extremely open with
widely scattered trees. Bird density is lower in these very open woods,
though birds such as Willow Warbler and Tree Pipit are typically

Fig. 8.1. Though widely
distributed on the mainland
of Europe, in Britain the
Crested Tit is confined to a
very discrete range within
Scotland. The population is
divided approximately
equally between pine
plantations and fragments
of ancient pine forest.
Chris Rose.

more abundant where the canopy is open. Secondly, the development of the field and shrub layers varies considerably. In part, this may be related to overgrazing and to canopy openness (see below). A dense understorey of juniper and birch offers nest sites for birds such as Dunnock, Long-tailed Tit and thrushes, while a rich field layer of low ericaceous and *Vaccinium* shrubs provides food for woodland grouse. The third factor is the maturity of the forest. The older, least disturbed stands have the greatest amounts of dead wood. Overall bird density appears to be higher in those areas with large amounts of dead wood (Hill *et al.*, 1990). Crested Tits need soft standing dead wood in the form of stumps in which to excavate nest holes. These tits reach their highest densities in the most natural pinewoods, though they also nest in pine plantations. In view of the relatively large extent of maturing plantations, these may soon support the majority of Scottish Crested Tits (Petty & Avery, 1990).

There is scope for increasing the amount of natural pinewood habitat in Scotland by the improvement of regeneration in the more open stands and by encouraging pine to spread onto moorland adjacent to existing woods (Williams & Green, 1993). The main hindrance to this is the high numbers of grazing animals, especially deer, which not only inhibit pine regeneration but also reduce the heather, bilberry, cowberry and juniper shrubs that form the field and shrub layers. The structure of much existing pine woodland has been greatly modified in this way. Apart from having a deleterious effect on the bird communities, extreme grazing pressure threatens the long-term survival of the wood by the prevention of regeneration. Conservation bodies, supported by the Forestry Authority and Forest Enterprise, are now aiming to stimulate regeneration in some native pinewoods and on adjacent land by the control of grazing. Effective deer fencing is expensive and, in any case, it merely shifts the problem elsewhere. Fencing also constitutes a collision hazard for Capercaillie and Black Grouse (Catt *et al.*, 1994). The only permanent solution is to reduce the numbers of animals and to maintain them at levels which are compatible with a diverse forest structure. There are, however, immense problems to be overcome in order to achieve this over sufficiently large areas.

Assuming the deer can be brought under control, there are two basic approaches that would maintain the value of these forests to wildlife. In some large tracts of ancient pinewood, and in some new areas of natural regeneration, a policy of non-intervention would seem highly desirable. This would be the best way to develop a natural forest structure with a patchy canopy cover, a varied shrub layer, many mature trees, plenty of dead wood, and with regeneration occurring in treefall gaps. Indeed, the Royal Society for the Protection of Birds has adopted this policy in the more natural pinewoods within Abernethy Forest (Hunt, 1990). The second approach would be low intensity management, involving thinning and periodic felling of small groups of trees. This would be most appropriate where stands were

Fig. 8.2. The red deer is a natural denizen of the Caledonian Scots pine forest but numbers are currently so high in many parts of the Scottish Highlands that the structure of the forests is impoverished and the prospects for regeneration of new areas of forest are severely restricted. W.S. Paton.

clearly lacking a natural structure. The aim would be to create canopy openings to stimulate vigorous growth of dwarf shrubs, juniper, and birch beneath the pine. Scattered trees could be left to mature and eventually become sources of dead wood. Such a minimum intervention policy, coupled with natural regeneration, would give stability to the overall appearance of the forest. In some cases, this type of management could be seen as a short-cut to the creation of a more natural woodland structure and, after a period of management, it may be deemed best to revert to non-intervention. Thinning at various intensities, combined with felling small groups of trees, can also be used to give existing Scots pine plantations some of the characteristics of natural pine woodland. This treatment may be appropriate in those places, such as nature reserves, where timber production is of secondary importance to conservation. Pole-stage plantations managed primarily for timber production can never look like old natural Scots pine woodland.

Red deer inhabited the original pine forests and it is entirely appropriate for them to live in today's pinewoods. Nothing is known about the population levels of grazing animals that existed in the primeval pine, but wolves and lynx probably maintained deer numbers at well below their present levels. In the absence of predators, there will need to be a continual commitment to the management of deer populations.

Yew woodland

The yew has two main centres of natural distribution in Britain: northern England, especially on the Carboniferous limestone, and the chalk of the south. Most yew woods are rather small but there are fine examples on the South and North Downs. Much of what is known of the birds of yew woodland comes from a study of the famous yew wood at Kingley Vale, Sussex (Williamson & Williamson, 1973; Williamson, 1978).

Many of the trees in southern yew woods are several hundred years old. They can form almost pure stands, devoid of any field or shrub layer, and these places are poor habitats for breeding birds. However, yew often grows in a mixture with broadleaved shrubs and trees and here the bird community can be much richer. At Kingley Vale, relatively few species were found in pure young and middle-aged yew. The greatest numbers of species bred in a yew–oak mixture with areas of scrub, and a mature yew–ash mixture also carried a fairly rich bird community. The commonest species at Kingley Vale were Chaffinch and Robin which, together with Woodpigeon, Blackbird,

Wren and Dunnock, made up half the total number of pairs. Goldcrest and Coal Tit are characteristic of yew but, in Britain, they do not seem to reach such high densities as in spruce and pine. Wren, Dunnock, Nightingale and warblers avoid stands of pure yew but may hold territory either at the edges, where the yew gives way to grassland, or in scrub or woodland where the yew is mixed with other species. Batten (1976) reports the results of a census of an Irish yew wood where, in contrast to Kingley Vale, Goldcrest was the commonest species.

Autumn and winter may see congregations of thrushes feeding on yew berries. In late summer, quite large flocks of Mistle Thrushes may descend on the yews and stay well into the winter. The berry crops on some of the trees may be defended by Mistle Thrushes and not eaten until late in the winter. In October, large numbers of Redwings and Fieldfares join the local thrushes and they remain the most abundant species until their numbers decline in mid-winter as the berry stocks are depleted. Thereafter, Blackbirds are the commonest thrushes. The berry-feeding thrushes are the most conspicuous birds in winter but a wide range of other woodland birds feeds in yew woods at this time.

Conifer plantations: from planting to felling

Large changes in bird communities of plantations occur between planting and harvesting. These are driven by the rapidly changing structure of the forest and the development of associated food resources. The intention here is not to provide a full list of the species that use the different stages of growth, but rather to give an impression of the transformation undergone by the bird life as the trees grow. There is, of course, much variation according to tree species, soil conditions and climate, to which I return below. Nevertheless, some general patterns can be discerned (Fig. 8.3). While individual trees grow in stature throughout the forest cycle, up to a height of perhaps 20–25 m at a harvesting age of about 50 years, much of the turnover in bird populations is linked with changes that occur in the first 20 or so years. Thereafter, the bird community changes rather little. A critical event is the closing of the tree canopy which occurs in the pre-thicket stage, typically soon after 10 years in many lowland forests, but usually at 12 to 18 years in the uplands. This rapidly shades out the vegetation close to the ground, including most non-crop trees and shrubs. Canopy-closure marks a watershed between species associated with the open phases of the forest and those of the mature plantation. A few species are mainly associated with the canopy-closure period itself. In Thetford Forest, these species include Golden Pheasant, Turtle Dove and Willow Warbler. Summer visitors (notably Nightjar, Tree Pipit, Whinchat and warblers) are largely restricted to the younger stands, while seed-eaters (mainly Siskin and Crossbill) are mostly confined to the older stands. Despite the huge shift in bird species composition, there appears to be an underlying trend in many forests to an overall increase in the numbers of breeding species, and in total bird density, from planting to harvesting. But in some forests, such as Thetford, the thicket stage holds more species than later stages (P. Donald; R. Hoblyn, personal communication).

There has been considerable debate about whether the bird communities of young, new afforestation, on open moor or heath, differ from those on restocks within existing forests. Some differences are to be expected simply because some birds of the original open habitat may continue to live within young afforestation for a short period; conversely, woodland birds may more rapidly colonise felled and replanted sites that are in close proximity to established trees. In upland Wales, restocks can carry more varied bird communities than afforestation (Bamford, 1985; Bibby *et al.*, 1985), but recent research

Fig. 8.3. Some changes in conifer plantations established on heath or moorland in relation to their stage of growth. Changes in vegetation structure (upper), availability of major food resources (mid), and abundance of some commoner birds (lower) are shown. Trends in vegetation structure and bird abundance simply indicate when each is at a maximum and minimum level. Bird abundances are not comparable between species. The maximum amount of a food resource is shown by the largest circle, the least amount by the smallest circle. Based on Lack & Lack (1951), Moss *et al.* (1979), Ratcliffe & Petty (1986), Sykes *et al.* (1989).

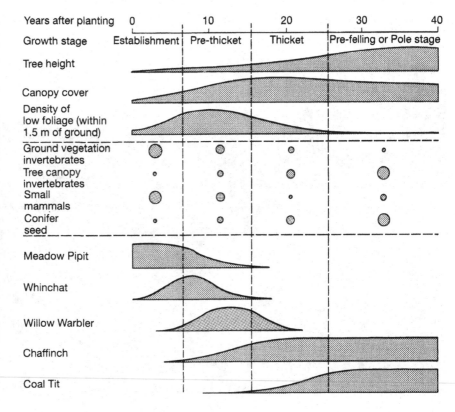

in northern Britain has not detected any such differences in overall bird numbers or species diversity (S.J. Petty, personal communication). It appears that, compared with afforestation, Welsh restocks generally develop more varied field and shrub layers, with much scrub invasion. As a result, these restocks can support higher densities of scrub species such as Dunnock, Whitethroat and Garden Warbler. In northern Britain, where deer numbers are higher and soils tend to be less productive, the restocks are less likely to develop such varied vegetation. A lowland example of differences between first- and second-generation forests is given by the Tree Pipit in Thetford Forest. This bird was uncommon on the recently afforested Breckland heaths (Lack & Lack, 1951), yet it is abundant on restocks in these same forests today. Yellowhammer, on the other hand, appears to be a common bird in lowland forests, both in afforestation and restocks. Short-eared Owls will nest at high densities on upland restocks, as well as on afforestation, in years of high vole numbers. Hen Harriers, however, relatively rarely nest on restocks (Bibby & Etheridge, 1993). Other raptors, including Sparrowhawk, Goshawk, Kestrel, Barn Owl

Fig. 8.4. A young restock in Thetford Forest, with a retention of old pine alongside the track in the middle distance. Clear-fells in Thetford Forest are colonised almost immediately after felling by Woodlarks and Nightjars. The vegetation in this restock is now almost too dense for Woodlarks, though a pair did nest close to this track in the year that this photograph was taken. This restock will, however, remain suitable habitat for nesting Nightjars for several years to come. Ron Hoblyn.

and Long-eared Owl, will regularly hunt over restocks if suitable nest sites are available nearby.

By the mid-thicket stage, the bird species associated with scrub are disappearing, and are rapidly being overtaken by birds that feed in the high canopy and which do not rely on a dense understorey for food, cover or nest sites. Some of these birds, like the Chaffinch, may have been present since the late-establishment stage but their numbers increase greatly during the thicket stage. Working in the 1950s, Gibb (1960) found that Woodpigeons, Coal Tits and Chaffinches were the commonest breeding species in 20–30 year-old stands of pine in Thetford Forest. Next most abundant were Pheasant and Wren, followed by Willow Tit, Long-tailed Tit, Goldcrest and Treecreeper. Only where nest boxes were present did Blue and Great Tit join this list of relatively common species, but they never approached the densities of Coal Tit, which is by far the most numerous tit in British conifer plantations. Blackbird and Song Thrush were found to be surprisingly scarce in Thetford Forest, but in other conifer plantations, for instance in the Forest of Dean, they are common. It is worth mentioning that Willow Tit is now absent, or at least extremely rare, in the pine stands of Thetford Forest. While it is a typical species of conifers in Fennoscandia, its status in British conifers is uncertain.

The bird communities of the oldest forest stands are often dominated by Chaffinches, Goldcrests and Coal Tits. Some pre-felling stands also carry high numbers of Wrens and Robins but in many conifer woods these birds are commoner in pre-thicket and thicket stages respectively. As stands mature, so the amount of available conifer seed increases (Fig. 8.3). It is not surprising, therefore, that Siskins and Crossbills are usually among the last breeding birds to colonise plantations. Both species will, in fact, breed in stands from pre-thicket onwards but their numbers are generally greatest in pole-stage stands (S.J. Petty, personal communication).

Several birds of prey have adapted to thicket stage and pre-felling plantations. The Goshawk has increased considerably in number in British coniferous forests during the past decade, though it is still a rare bird. By contrast, Long-eared Owl is locally fairly common, for example in Thetford Forest, but it is absent (or under-recorded) from some other forests with apparently suitable habitat. Merlins have started to nest in old crows' nests in conifer trees at the edges of some upland plantations (Parr, 1991; Little & Davison, 1992). The two most widespread birds of prey in conifer plantations are Tawny Owl and Sparrowhawk. In conifer plantations in southern Scotland, Sparrowhawks prefer young dense stands of 20–35 years' growth (Newton, 1991). These stands are occupied more frequently, and breeding within them is more successful, than in older stands. It is likely that the late thicket and early pole-stage stands are better habitats for Sparrowhawks as a result of greater food availability, perhaps because the birds they prey upon are in some way easier to catch there, rather than simply being more abundant. It is also possible that nests in the preferred habitat are more sheltered from inclement weather and predators.

There can be large variation in the structure of pre-felling stands, even over quite small distances. In Thetford Forest, for example, the

Fig. 8.5. Conifer forests are the main habitat of the Goshawk in Britain. This species has become firmly established in Britain over the last twenty or so years. It will probably continue to expand its range and population, taking advantage of the large forests that have been created this century, particularly in the uplands. Chris Rose.

ground beneath pine stands in one part of the forest may be entirely bare of vegetation yet in another stand of the same age, the forest floor can be carpeted with bramble and scattered elder. This is an effect of soil type and it has a considerable impact on the bird communities. The presence of a shrub layer usually leads to a higher overall density of breeding birds and species often absent from mature conifer stands, such as Blackcap, may breed there.

The dominant tree species also affects the bird communities of plantations. In general, pine carries fewer birds than spruce (Newton, 1986*b*). This may be because spruce offers greater amounts of insect food, or possibly better shelter for nests. There have, however, been few comparisons between the different species of spruce or of pine. These would be valuable in view of the changing patterns in tree planting described in Chapter 1: the area of Sitka spruce is increasing, while Norway spruce is diminishing, and Corsican pine is replacing Scots pine in some lowland areas. In Thetford Forest, Gibb (1960) found that Coal Tits, Blue Tits and Willow Tits were more numerous in Scots than Corsican pine; this applied both in summer and winter.

In winter, many of the breeding birds desert the conifer plantations. The youngest stages can seem particularly devoid of birds. In Thetford Forest, Meadow Pipit is the commonest wintering species in the young pine restocks, even though it does not breed there. In thicket and pre-felling stands, the bird communities may be more similar to those of the summer, though some birds, such as Chaffinch and Robin, may be considerably reduced in numbers. In Thetford Forest, Gibb (1960) found that the species most commonly using pine plantations in winter were those that were most abundant as breeding birds (see above). The main difference in winter was that Pheasants, Wood-pigeons and Chaffinches used the pine mainly for roosting rather than feeding. Thicket-stage conifers are widely used by several finches for roosting. In March and April, there can be a major influx into pinewoods of Greenfinches, Goldfinches, Chaffinches, Bramblings and Siskins to feed on falling pine seed. Recently, large flocks of finches have also been reported feeding on Sitka spruce seed (Shaw & Livingstone, 1991). Cone production in conifers varies strikingly from year to year and this has a big effect on the numbers of Crossbills and Siskins that use thicket and pole-stage stands in winter. For example, between 1974 and 1986 exceptionally large numbers of Crossbills and Siskins were present in Kielder Forest in those years when there were large crops of Norway spruce cones (Petty & Avery, 1990). In Finland it has long been known that the abundance of the spruce cone crop determines the density of Crossbills that breed in any particular year (Reinikainen, 1937).

Fig. 8.6. Recent changes in the distribution of the Siskin in Britain and Ireland as shown by the two atlases of breeding birds (Sharrock, 1976; Gibbons *et al.*, 1993). The dramatic expansion of range is a consequence of increased availability of habitat as conifer plantations, especially those of Sitka spruce, have matured (Gibbons & Gates, 1994).

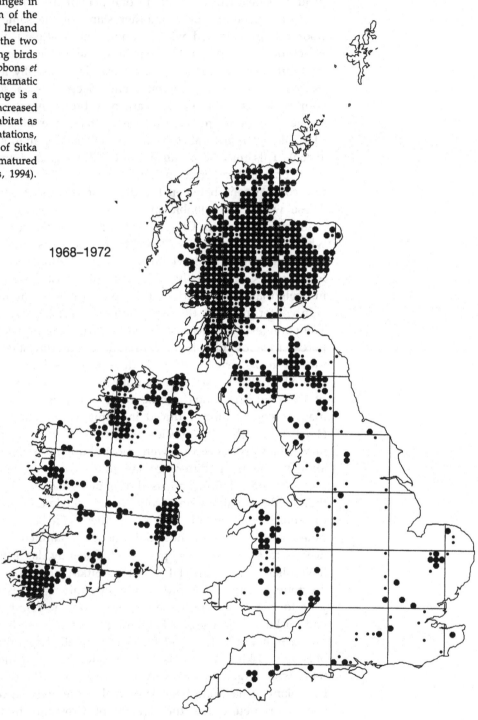

1968–1972

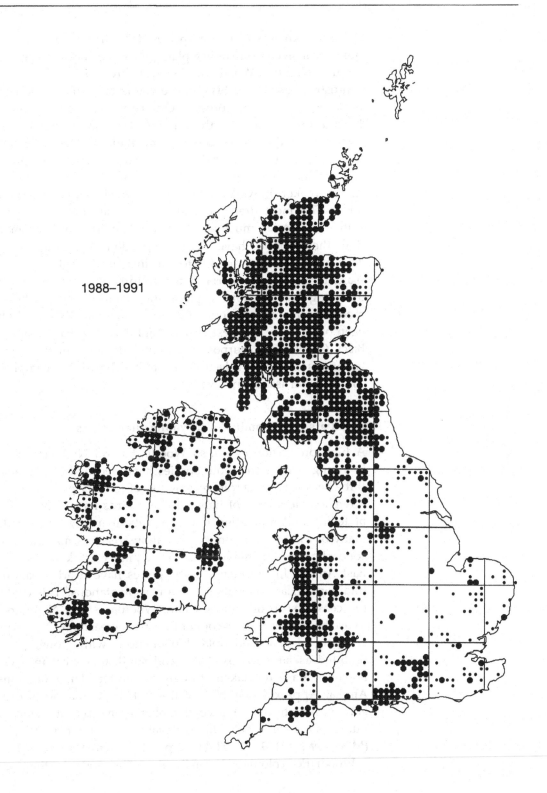

1988–1991

Ten of our less common species depend to a large extent, though none exclusively, on conifer plantations: Goshawk, Capercaillie, Black Grouse, Nightjar, Woodlark, Firecrest, Crested Tit, Common Crossbill, Scottish Crossbill and Siskin. The status of all these species in Britain is closely linked with forestry. One could add the introduced Golden Pheasant to this list, for the thriving Breckland population is strongly associated with middle-aged pine plantations. There is also a case for including Short-eared Owl because young plantations are one of its two primary breeding habitats in Britain. The case for Hen Harrier is rather weaker, however, because it is largely confined to new afforest-ation. The most dramatic examples of the effect of forestry on the status of species must be Common Crossbill and Siskin (Fig. 8.6). Both these species have hugely expanded their ranges over the last 20 years in direct response to the increased availability of maturing plantations in many parts of the country (Gibbons *et al.*, 1993). The following two sections explore the detailed relationships between modern forestry and four of these scarcer bird species which have been particularly well researched: Nightjar, Woodlark, Capercaillie and Black Grouse. The purpose is to illustrate some of the subtleties and variations of habitat selection and to emphasise the complexity of the conservation issues involved.

Nightjars and Woodlarks in young plantations

A high proportion of British Woodlarks and Nightjars is now found in young plantations. Both species have declined over the last two to three decades but their increasing use of forestry plantations represents a remarkable switch of habitat. These birds have exploited the phase of felling and replanting that came about in the 1970s and 1980s as many of the pre-war conifer plantations reached maturity. The largest concentrations of these birds are in pine restocks on sandy soils in East Anglia. In Breckland, both species have nearly abandoned their traditional heath habitats in favour of plantations. This contrasts with the coastal strip of Suffolk where Woodlarks and Nightjars continue to use heaths, as well as conifer forests. Increasingly, Nightjars are also breeding in restocked upland plantations with growing populations reported as far apart as Wales and south-west Scotland. Woodlarks, however, have not taken strongly to conifer plantations outside East Anglia. Approximately 50% of the British population of each species is now thought to live in conifer plantations (recent estimates of total numbers are some 3400 Nightjar pairs and at least 500 Woodlark pairs (Morris *et al.*, 1995; R. Hoblyn, personal communication).

Woodlarks colonise clear-fells almost as soon as they are created but rarely use them beyond 5 years after restocking. The birds strongly favour stands where the trees are 1 to 2 years old (Fig. 8.7a). The reason is that these stands are the only ones to provide suitable feeding habitat (Bowden, 1990; Bowden & Hoblyn, 1990). The birds require substantial areas of bare ground and short vegetation (<5 cm), with freely available perches. Woodlarks select feeding-sites where short vegetation covers more than 50% of the area (Fig. 8.7b). The birds feed by walking and taking invertebrates from the ground or low vegetation. They can feed efficiently in this way only when the vegetation is very short. Woodlarks shun stands older than 5 years not because they are deterred by the growth of the trees, but because the ground vegetation becomes too dense.

While the Woodlark has adapted to new habitats created by forestry, it is vulnerable because it depends on such a narrow 'window of opportunity'. Indeed, during the next decade or two, there will be far less felling and restocking in the East Anglian plantations than has been the case since the 1970s. Consequently, there will be far less suitable habitat available for the birds and the population is predicted to crash (Bowden & Hoblyn, 1990). To help overcome this bottle-neck period, trials are being conducted to assess whether herbicides or ploughing could be used to extend the suitability of stands. At the time of writing, however, the Woodlark population of Thetford Forest is increasing strongly and showing no sign of being limited by habitat availability (R. Hoblyn personal communication).

Nightjars are not quite so demanding in that they occupy a wider

Fig. 8.7. Use of pine plantations by Woodlarks in Thetford Forest. (a) Density of breeding Woodlarks in relation to the age of the trees. (b) The percentage of cover of short vegetation, less than 5 cm tall, at feeding sites (upper) compared with randomly selected study sites in the same area (lower); each site consisted of an area measuring 4 × 6 m. From Bowden & Hoblyn (1990).

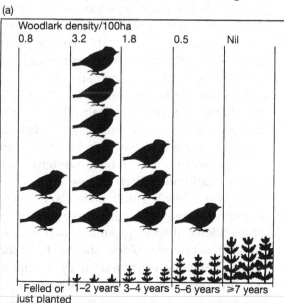

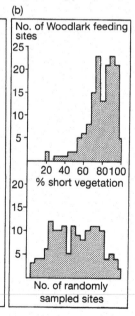

range of plantation ages than Woodlarks. Nightjars will use restocks throughout the establishment and pre-felling stages. In coastal Suffolk, Ravenscroft (1989) reported highest densities in 3 to 5 year-old pine but Nightjars will certainly use considerably older growth. In Thetford Forest and elsewhere, densities remain fairly high until at least 10 years, with birds continuing to use some stands for up to 15 years. Being an aerial feeder, the Nightjar is not as critically dependent on the vegetation structure for feeding as the Woodlark. Indeed, the bird may well feed up to several kilometres from its nest (Alexander & Cresswell, 1990). Nonetheless, it does need bare ground or sparse vegetation for nesting. Ravenscroft suggested that the use of herbicides to reduce the cover of bramble and bracken, and mechanical swiping to remove non-crop trees, might prolong the suitability of a stand for nesting Nightjars.

There is much to be learnt about why Nightjars and Woodlarks use plantations in some regions but not in others. It could be that the vegetation structure is sometimes unsuitable, even when the age of the plantation is 'correct'. Both species are traditionally associated with dry places, so perhaps some sites are too damp, particularly in the uplands.

Black Grouse and Capercaillie

Our two forest grouse have been in serious decline for some years, not just in Britain but throughout Europe and Fennoscandia. While forestry has offered a safety net to the dwindling populations of Nightjars and Woodlarks, this is not so clearly the case for Black Grouse and Capercaillie. Though not yet fully understood, the problems facing these splendid gamebirds are extremely complex and rather different for the two species.

The Capercaillie was reintroduced into eastern Scotland in the nineteenth century and it remains largely restricted to the central and eastern Highlands. The breeding productivity of the species is sensitive to high rainfall and this may be the reason why the Capercaillie has not spread into the rainier, western parts of Scotland. The more widely distributed Black Grouse shows no such adverse response to rain (Moss, 1986). Adult male Capercaillie live throughout the year in mature stands of conifers, especially Scots pine. The males gather at traditional display grounds (leks) in the spring, these are usually situated within the oldest stands. In Scotland, woodland containing leks is generally characterised by a semi-natural structure with old, 'granny' trees (Picozzi *et al.*, 1992). Scottish Capercaillie, however, are not restricted to old forest and some large leks occur in plantations.

In winter, females and subadult males may use middle-aged as well as old stands (Gjerde & Wegge, 1989). Black Grouse typically live at the interface of moorland and open woodland, and occupy a wider range of habitats than Capercaillie. There are large populations in upland regions of Scotland and northern England, with a smaller one in Wales. Their leks are usually situated on open ground close to birchwoods or young conifer plantations. Some leks are located within the woodland itself but the birds mostly use the woods for feeding. Black Grouse need this mosaic of different vegetation types to satisfy the needs of nesting, roosting and of feeding at different times of year (Parr & Watson, 1988; Cayford & Hope Jones, 1989; Baines, 1990).

Three general types of factors could lie behind the declines in forest grouse: habitat loss, deterioration of food resources, and predation. Taking each of these factors in turn, I shall start with habitat loss.

For Black Grouse the important habitat changes have been the improvement of rough grazing, drainage of wet flushes and mires

Fig. 8.8. A high proportion of the British Nightjar population now lives within commercial conifer plantations. Chris Rose.

which are important feeding sites, and clearance of birch scrub. The advent of large areas of conifer plantation may have actually aided the species in some areas but this effect is temporary for the birds cease to use plantations once the canopy closes. Maintenance of a continuity of suitable habitat within upland forests will be an important component of Black Grouse conservation in the future. Recommendations for achieving this are made by Cayford (1993). The main habitat loss suffered by Capercaillie has been that of old pine forest. Fortunately the importance of the surviving fragments is now recognised, but much has been lost, even in the last 50 years. The size of old forest patches is crucial to Capercaillie. The number of cocks per lek, as well as the number of leks, increases with the size of patch. In a Norwegian study, the smallest occupied patch was 48 ha, while less than half the patches examined between 50 ha and 100 ha held leks. To be sure of holding a lek, a patch had to be at least one square kilometre (Rolstad & Wegge, 1987). The survival of Capercaillie living in areas where the old forest had been highly fragmented was also lower than that of birds living in more continuous habitat

Fig. 8.9. Native Scots pine forest with large mature trees and a luxuriant field layer; these are important attributes of good quality Capercaillie habitat. Nick Picozzi.

(Gjerde & Wegge, 1989). Recommendations for Capercaillie habitat management are given by Moss & Picozzi (1994).

The second important factor is the quality of feeding habitat. Both grouse rely on a prolific ground vegetation as a source of vegetable and invertebrate food. Overgrazing by deer and sheep in areas of the Highlands appears to have reduced bilberry to a point where there is a serious reduction in the amount of associated lepidopteran caterpillars which are a vital food for the chicks of forest grouse (Baines & Sage, 1992). A similar effect probably occurs in many pine plantations where the dwarf shrubs are reduced due to shading by the closely spaced trees. A practical solution where conservation bodies own plantations is to undertake a heavy thinning. On moorland, drainage may also have reduced the food-plants of sawfly larvae which are a food of Black Grouse chicks (Baines, 1991).

The final factor is predation. This is a complicated subject which has not been thoroughly researched in British forests, so much of what I have to say about possible links between predators and grouse is speculative. Numbers of predators have probably increased as keepering has become generally less intense during the twentieth century. In Scandinavia, the breeding success of forest grouse is strongly cyclical and it parallels natural cycles in vole numbers. When voles are numerous, predation pressure eases on the grouse, but when vole densities plummet, the predators turn more of their attentions to the grouse (Angelstam et al., 1984). There is no evidence that voles and grouse act as alternative food sources in this way in Britain. Neither is there evidence, though it remains a possibility, that modern forestry has increased the overall numbers of corvids and foxes, thus contributing to generally poorer breeding success of forest grouse, as suggested for Finnish forests (Henttonen, 1989).

Since they are often unkeepered, forestry plantations may act as a refuge for predators, as well as providing nest sites for corvids. The periodically high densities of voles in clear-fells certainly provide a huge food resource for predators. One may, therefore, expect high rates of predation on birds, such as Black Grouse, nesting on open land at the edges of plantations. There have been few relevant studies in Britain but Avery et al. (1989) looked at predation rates on moorland in Sutherland using dummy nests containing chicken eggs. They found that losses of eggs decreased linearly with distance from the edges of plantations but concluded that this effect was due to spatial differences in vegetation type on the moorland and not to the proximity of the plantation. This study would be worth repeating elsewhere in upland Britain. It is also possible that the level of predation within the forest itself may be related to the extent to which the forest is fragmented into a patchwork of different growth stages. Predators attracted to the clear-fells may inflict losses on the Capercaillie nesting close to clear-fells. Work in Sweden, again using dummy nests placed on the ground, but this time within the forest itself, showed that losses of eggs were higher at the edge of the forest, particularly in the outer 50 m, than in the interior (Andrén & Angelstam, 1988). However, this research was carried out at the interface of farmland and forest, and the results may not be applicable to edges between clear-fells and mature forest (Rudnicky & Hunter, 1993).

Improving the conifer forests for birds

The new conifer forests support huge numbers of birds and already several species in Britain are largely dependent on them. Their conservation value is more widely appreciated than was the case 10 years ago but there is still some way to go before their full conservation potential is realised. Much has been learnt about the ecology of these forests in recent years and the prospects are excellent for creating second-generation forests that are more varied and appealing to both

Fig. 8.10. Male Capercaillie in its forest habitat. Nick Picozzi.

people and wildlife. The Forestry Commission has fully recognised this opportunity and enormous advances have been made in making wildlife habitat management an integral part of forestry operations.

As far as scarce species are concerned, it is essential to understand their requirements so that appropriate conservation measures can be defined (for a summary see Batten *et al.*, 1990). In addition, various modifications can be made to the design and management of conifer forests that will lead to generally richer bird communities. The five broad approaches summarised below are not alternatives, but ideally should be adopted simultaneously. These ideas are not my own; for more detail see Ratcliffe & Petty (1986), Avery & Leslie (1990), Petty & Avery (1990). The principles discussed in Chapter 4 are also highly relevant to understanding the effects of forest design and management on bird populations. Special 'wildlife zones' can be extremely valuable but the concept of conservation management should not be limited to uncropped areas. There are opportunities within the conifer stands themselves. Retentions and systems other than clear-felling are especially exciting prospects for creating new types of habitats within some forests.

1 *Uncropped land* It is highly desirable to create networks of semi-permanent wildlife habitat within forests, perhaps covering at least 10% of the total forest area. These would encompass linear features including rides, forest roads, wayleaves and watercourses, all of which could be used to link patches of broadleaved woodland and unplanted areas such as bogs and crags. This uncropped land could be maintained so as to provide habitats for many species that were scarce or absent within the crop itself. We must hope that the days have gone when conifers were planted without hesitation up to the edge of all streams. Watercourses offer ideal sites for the creation of mosaics of open space and broadleaved trees.

2 *Forest restructuring* As described in Chapter 1, restructuring of the first-generation plantations gives the opportunity to move towards forests that contain a greater variety of age classes. This will enable forests to support a broader range of bird communities. During restructuring it is also possible to create a greater diversity of coupe sizes. Some species will benefit from small coupes, where edge effects will be high, and others from large coupes. Within large forests, the best strategy is probably to aim for a variety of coupe sizes.

3 *Tree species* The value of small patches of non-crop trees may be out of all proportion to the area they cover. Introduction of some broadleaved trees where they do not already exist, either as small

patches or interspersed in the conifers, is especially desirable. These trees may offer important feeding sites for many birds as well as nest sites, particularly for hole-nesters. In upland Welsh plantations, Bibby et al. (1989) found that 11 out of 22 species of breeding birds preferred areas with broadleaved trees and that the overall density of these species increased in relation to the area of broadleaved trees. They also showed that widely scattered broadleaved trees had a greater positive effect on bird numbers than did large blocks of broadleaved trees. Broadleaved patches in lowland conifer plantations are also selected by some birds (Williamson, 1972b). Some non-crop trees that have naturally colonised young plantations can be encouraged to grow on within the crop. Maintaining a small proportion of European conifers, say 5 to 10% of Scots pine, Norway spruce or larch can be beneficial to seed-eating birds.

4 *Retentions and dead wood* Retaining patches of forest to grow on well beyond natural rotations can be of great conservation value. Many such areas would still be treated as part of the harvested crop – their rotation would simply be longer, perhaps twice as long as most of the crop. Others can be managed by continuous-cover systems or even left indefinitely. Retentions could have an obvious benefit to Capercaillie and Goshawks. The density and richness of songbirds could also be greater in such retentions, partly because certain types of nest sites, especially holes, are more numerous (Currie & Bamford, 1982). The idea of creating networks of old-growth patches within upland conifer forests is worthy of serious consideration (Peterken et al., 1992). With the exception of areas of windthrow in upland forests, dead wood is a rare commodity in plantations, but the adoption of retentions would help to rectify this. Another, complementary, approach is to leave snags (standing dead trees) within forest compartments. The current practice in Thetford Forest is to leave snags in clear-fells at a density of approximately one per hectare. The value of snags to hole-nesting birds has been firmly established in North American and Scandinavian conifer forests (e.g. Dickson et al., 1983). This subject would be worth exploring in British forests to assess the densities of snags that would be needed to make an impact on the density of hole-nesting birds.

5 *Systems other than clear-felling* Coniferous forestry in Britain is almost exclusively based on clear-felling systems. This situation exists partly for economic, partly for technical and partly for traditional reasons. Clear-felling strongly benefits some birds, for example Nightjars and Woodlarks, so one would wish to see it continue in many areas. In lowland broadleaved woods, however, there has been a recent

Fig. 8.11. Leaving moderate numbers of snags, such as dead birch trees, both within the growing conifer crop and in clear-fells, may be beneficial to hole-nesters. Chris Rose.

move towards smaller-scale operations, or group-felling (Chapter 1). Trials with similar systems have been advocated for some upland forests (Avery & Leslie, 1990). Wherever possible it would be desirable to adopt natural regeneration. Effective deer control would be vital for success. The resulting forest would be a fine-scale mosaic of patches containing trees of different ages. This structure would certainly mimic a more natural forest than most clear-felling systems, especially if some of the patches were allowed to grow on indefinitely. Exactly what sort of bird communities might develop is unknown but it would be fascinating to find out.

9

Woodland in a changing countryside

The countryside of Britain encapsulates a wide variety of scenery and vegetation. Most of these familiar landscapes have evolved over hundreds, even thousands, of years as man has reshaped the land to suit his needs within the natural limits of soils, topography and climate. The view from the Cornish room where I write these words could not be a more powerful reminder that while the countryside has ancient origins, it can never be static. It continues to be adapted and altered but, when neglected, it throws up unplanned wildness. The full pattern of tiny fields laid out in some former age across the slopes of this Penwith valley can now hardly be traced. Over the last 50 years, fields have steadily been abandoned. Many were once plots for growing early spring flowers. Isolated patches of grass remain, each demarcated by a stone hedge from the surrounding bracken, bramble, gorse and blackthorn. The mature sycamore and ash flanking the stream in this valley provide one of the most westerly British outposts for woodland birds such as Green and Great Spotted Woodpecker, Jay, Goldcrest, Treecreeper and Nuthatch. Dozens of nesting Rooks crowd into these trees each spring. If the scrub grows unabated, which for the most part seems likely, these birds will eventually be able to exploit a new woodland habitat on the valley sides. Already sycamore and ash saplings are springing up in many fields, and one landowner has cleared two small patches of scrub to plant trees, some native, some not. In the meantime, other species of birds benefit from the scrub. Dunnocks and Wrens are probably the commonest breeding birds in the scrub, which is also attractive to large numbers of breeding and migrating warblers. In the span of just one century, the bird populations in this western valley must have responded profoundly to the changing land-use. The present mosaic of grassland, scrub and woodland habitats probably supports a more diverse bird life than at any time this century. The long-term trend towards woodland is more likely to impoverish than enrichen the future bird life of the valley.

Few areas of Britain may be witnessing such an extreme transformation as this Cornish valley, but many changes are underway in the countryside that have implications for woodland and its birds. Recent trends in woodland are summarised in Table 9.1. Some of these changes are widespread, others are mainly confined to nature reserves and other specially protected areas. On balance, these changes have been broadly beneficial for woodland birds. The most obvious deleterious change is the continued spread and increase of deer (Chapters 7 and 8). In this chapter, I discuss several issues concerning the future of British woodland and its bird life that have not been dealt with in earlier chapters. An important role of ecology is to predict how changes in land-use will affect the numbers and distribution of plants and animals. This chapter aims to assess what we know about the effects on birds of current changes in woodland and its management, and to highlight areas needing further research.

Damage to forests by air pollution

Air pollution has a largely unknown, but possibly insidious, effect on the ecology of British woodland. Local damage to forests as a result of air pollution has long occurred in industrial countries. High emissions of sulphur dioxide can cause trees in the vicinity to suffer defoliation, reduced growth and low survival (Prinz, 1987; Krause, 1989). In the last two decades concern over the effects of air pollution has intensified as forest damage has emerged on an unprecedented scale in Europe. The most dramatic deterioration in the health of forests has been in parts of Germany, Poland, and the former Czechoslovakia. Forests of Norway spruce and silver fir at high altitudes have been the most severely damaged. The name given by Germans to this massive death of forests is *Waldsterben*. During the 1980s, surveys pointed to even more widespread declines in the health of several tree species in Europe, including Britain, as shown by symptoms such as crown-thinning and the yellowing of foliage. The problem is not confined to conifers; for example, there is some evidence of a deterioration in the health of beech trees in Britain (Ling *et al.*, 1993; Woodin & Farmer, 1993).

Controversy surrounds the causes of this recent extensive forest decline, in particular the exact role of air pollution remains uncertain although it is widely accepted that pollutants are implicated. Air pollution may damage forests in two broad ways (Prinz, 1987; Krause, 1989). First, acid rain (i.e. sulphur and nitrogen deposited through precipitation in the form of acids) can have an indirect effect on trees through acidification of the soil, leading to loss of nutrients critical for plants and the mobilisation of harmful metals. Magnesium

Table 9.1. *A summary of recent attitude, policy and environmental changes with significant implications for woodland conservation in Britain. Key references in parentheses. See Dudley (1992) for a general review of recent changes in temperate forests.*

1. The importance of ancient woodland, both as refuges for scarce plants and invertebrates, and as sites of cultural significance, is widely accepted (Rackham, 1980; Spencer & Kirby, 1992; Peterken, 1993). An increasing body of ancient woodland is protected within reserves.

2. Traditional management (coppicing and wood-pasture), and other systems that create similar conditions, is now widely accepted as important for the maintenance of populations of many scarcer woodland plants and animals, especially within ancient woods (Chapters 6 and 7). An increasing body of management guidance is available for reserve managers (reviews by Fuller & Warren, 1991; Kirby, 1992; Fuller & Peterken, 1994).

3. Other management systems are now recognised as producing habitats and communities that complement those of the traditional management systems. High forest systems have a valid role in many reserves. Establishment of *natural* or *old-growth* woodland through non-intervention is gaining increasing support as an objective for some reserves (Peterken, 1991; Fuller & Peterken, 1994).

4. The importance of broadleaved woodland, both for conservation and landscape reasons, is recognised in the lowlands, and the presumption for most lowland planting and restocking is now in favour of broadleaves. There are even reasonable prospects for restoring broadleaves on some ancient sites which were planted with conifers in recent decades.

5. Creation of new broadleaved woodland in the lowlands has accelerated. Financial initiatives (Farm Woodland Premium Scheme, set-aside schemes) exist to encourage planting on farmland. The concept of Community Forests on urban fringes was developed in the 1980s, though individual schemes are still in their infancy.

6. The Forestry Commission has assumed a statutory responsibility to strive for a balance between productive forestry and the conservation of landscape and wildlife (Wildlife and Countryside (Amendment) Act 1985). Wide integration of conservation into commercial forests has been achieved (Forestry Commission, 1990; see also Chapter 8).

7. Restructuring of first-generation conifer forests has given a major opportunity to create habitats for wildlife and to diversify their age structure (Chapters 1 and 8).

8. Recreational usage of woodland and forest has intensified. Locally severe impacts on vegetation and animals have occurred (Anderson & Radford, 1992) but the increased passive use of woodland can only serve to justify conservation management aimed at enhancing wildlife.

9. Deer numbers have increased enormously throughout much of Britain but especially in the lowlands where they cause severe problems for forestry and conservation alike (Chapter 7).

10. Acidification and air pollution have become recognised as serious long-term problems that affect the nature conservation value of many semi-natural habitats in Britain, including woodland (Woodin & Farmer, 1993).

deficiency in some acid central European soils has been pinpointed as a critical factor (Roberts *et al.*, 1989). Secondly, ozone and acid rain may directly attack the trees. It is extremely difficult to distinguish between the effects of air pollution and those of natural stresses, such as drought, and of forest management practices that may have led to a long-term reduction of soil nutrients (Roberts *et al.*, 1989). There may be complex interactions between stresses created by pollutants and natural stresses; for example, drought may raise the susceptibility of trees to pollutants. There are multiple causes to this forest decline that may differ according to the region and the tree species involved.

In those parts of central Europe with pronounced forest decline, the breeding bird populations of forests are strongly affected by the severity of the damage. In the former Czechoslovakia, within any particular age of spruce forest, the density, diversity and number of species of breeding birds decreased with increasing severity of damage (Šťastny & Bejček, 1985; Flousek, 1989). The density of birds in stands that are in various states of ill-health is illustrated in Fig. 9.1. Overall changes in density, however, conceal differences in the responses among species to forest damage. Flousek (1989), for example, found that Goldcrest, Firecrest, Blackbird, Song Thrush, Wren, Coal Tit and Chaffinch decreased in numbers as the forest damage increased. In contrast, Dunnock, Robin and Willow Warbler numbers remained stable, but Tree Pipit and Meadow Pipit increased. The pipits may have benefited from the opening of the canopy that occurred in

Fig. 9.1. Estimated densities of birds (pairs/10 ha) in mountain forests in the Czech Republic in relation to forest damage. A – Norway spruce, 40–80 years old, censused by territory mapping (Šťastny & Bejček, 1985). B – Mainly spruce forest, 60–140 years old, censused by point counts (Flousek, 1989). C – Same as B, but censused by line transects (Flousek, 1989). All these studies were made in the 1980s and B and C are based on mean counts over 4 years.

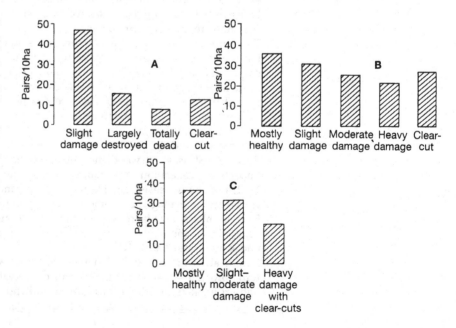

damaged stands. Other species may exploit the large amounts of dead wood, for instance Three-toed Woodpecker which often feeds by removing bark from dead spruces to reach the invertebrates living beneath. The benefits gained by these species are generally short-lived, however, because the most damaged stands are often clear-felled. These areas, even shortly after felling, can actually hold higher densities of birds than the most heavily damaged forest. The large-scale clearance of dead and dying forest has profoundly altered the bird populations of some regions. In south-west Germany, for example, regional populations of several species have risen substantially in response to the creation of clear-fells, some of which have rapidly established a dense bush layer (Hölzinger & Kroymann, 1984). Among these species are Black Grouse, Nightjar, Red-backed Shrike, White-throat and Grasshopper Warbler. These gains must, of course, be set against the losses of mature forest habitat which many other birds, especially hole-nesters, depend upon.

Various studies of tits reveal how damage to conifer forests can lead to reductions in the populations of insectivorous birds that feed in the foliage. Progressive thinning of needles in the spruce canopy almost certainly reduces the amount of invertebrate food available to birds. The abundance of spiders, for example, is known to be greater on branches with high densities of needles than on needle-thin branches (Gunnarsson, 1990). Massive declines of Coal Tits in spruce forests of the Harz Mountains are thought to be a consequence of poor breeding success caused by this reduction of their invertebrate food (Zang, 1990). A different conclusion, however, was reached in another study of Coal Tits and Crested Tits in damaged German forests (Möckel, 1992). Here, there was no evidence that breeding success was greatly impaired, but it was suggested that food shortage in winter led to a lower survival of adult birds. The Crested Tit partly compensated for these losses by producing a higher proportion of second broods, and its population decline was not as severe as that of the Coal Tit. Canopy-thinning may also cause birds to become more vulnerable to predators. In Sweden, Hake (1991) found that Willow Tits that fed in pine with heavy loss of needles spent more time apparently scanning for predators, and less time handling prey, than did birds feeding in areas with relatively little needle loss.

Changes in the quality, as well as the quantity, of food also appear important. Working on Great Tits in The Netherlands, Drent & Woldendorp (1989) found that the proportion of breeding pairs that produced no eggs, or eggs of poor quality, increased during the 1980s. This trend was particularly marked in that part of their study area that had poor soils. The cause was calcium deficiency as a result of acid rain. The uptake of calcium by trees was impaired, with the consequence that caterpillars feeding on the foliage contained less calcium. Interestingly, despite this problem, the population of Great Tits in the poor forest was not declining at that time because it was maintained by a surplus of production from the rich forest. This nice example of a population sink (poor forest) and a population source (rich forest), illustrates that simply monitoring bird numbers will not necessarily reveal the existence of serious underlying environmental problems. There have been no studies on the effects of air pollution on woodland birds in Britain, but it is very likely that subtle effects such as those described by Drent & Woldendorp are already in operation.

Climate change

Other major changes to woodland could be imminent in the form of increases in atmospheric carbon dioxide and associated climate warming. Most scientists agree that the greenhouse effect is a real phenomenon (Intergovernmental Panel on Climate Change, 1990), but opinions differ as to the extent of warming that will occur and to its ecological implications. The ramifications for communities of plants and animals will be enormously complex and individual species will show many different responses (Root & Schneider, 1993). Any assessments of possible impacts on temperate forests and their wildlife are speculative but interesting.

The geographical ranges of several broadleaved trees in Britain are probably limited by temperature. Quite small increases in mean temperature may cause trees such as beech, small-leaved lime and field maple, to grow much better and to expand northwards in range (Cannell *et al.*, 1989). Similarly, the ranges of many birds in Britain are probably determined by climate in one way or another, and it is very likely that several woodland birds would spread northwards. Candidate species include Lesser Spotted Woodpecker, Nuthatch, Willow Tit and Marsh Tit. These species may be aided in a northward expansion not just by climatic warming *per se*, but also by a diversification of the northern oak and birchwoods. There could be more subtle effects of climate change on birds, acting through changes in plant phenology and chemistry. For instance, it is predicted that tree species, such as hawthorn, which come into leaf early in the year, will do so even earlier as the climate warms (Cannell, 1990). This may alter the availability of insect food in some habitats, perhaps leading to earlier breeding seasons for some species of birds, and to an increase in the numbers of broods they can produce in a single breeding season. Increases in carbon dioxide will probably alter the growth and chemical make-up of trees, which in turn, will affect insects in complex ways (Lindroth *et al.*, 1993).

Human recreation and the effects of disturbance

Recreational activities in woodland range from the innocuous to the intrusive. Walking represents the passive end of the spectrum, while use of off-road vehicles and adventure games lie firmly at the other. The majority of woods in Britain remain fairly undisturbed, but there are some local problems. The number of woods suffering major disturbances from intrusive recreation has definitely increased in the last decade. The number of people indulging in relatively passive recreation in woodland is also probably increasing. There is no evidence that these trends are having any general effect on woodland birds but there may well be local impacts.

There are several aspects of human disturbance on woodland birds. In the most serious cases, the structure of the habitat can be so altered that it is physically no longer suitable for some species. This has

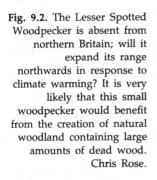

Fig. 9.2. The Lesser Spotted Woodpecker is absent from northern Britain; will it expand its range northwards in response to climate warming? It is very likely that this small woodpecker would benefit from the creation of natural woodland containing large amounts of dead wood. Chris Rose.

occurred in some woods where adventure games or motor cycling have devastated the field layer, and sometimes the shrub layer. Another potentially serious problem arises where dogs are frequently allowed to roam out of control. Though dogs may occasionally kill birds, the main problem is that repeated disturbance of nesting birds may cause nest desertion or increased exposure to predators. Of course humans can have the same effect, but they are less likely than dogs to penetrate dense vegetation. It cannot be assumed that passive recreation has no effects on birds. Bird species vary in their tolerance of people; some may be sensitive to even occasional disturbances.

Work in The Netherlands indicates that severe disturbance may reduce breeding bird numbers in woodland. One case study of a 50 ha wood reported declines in a large number of bird species over a 30-year period. These declines were partly attributed to increasing recreational pressure which damaged the field layer (Jansen & de Nie, 1986). Another study related the density of birds in urban woods to intensity of recreational use (van der Zande et al., 1984). A negative correlation between bird density and recreational pressure was taken as evidence of an effect of disturbance. Such evidence was obtained for eight out of 13 species: Woodpigeon, Turtle Dove, Wren, Song Thrush, Blackcap, Garden Warbler, Willow Warbler and Chiffchaff. These results, however, do not offer conclusive evidence that disturbance affects bird numbers because it is possible that some of the observations were a consequence of other factors not measured in the studies. Nonetheless, they reinforce the need to collect more information on the subject. In particular, it seems important not just to count birds in disturbed and undisturbed areas, but to measure their breeding success. It would also be worth studying whether there is competition between birds for undisturbed sites. It is not inconceivable that disturbed places, for example close to heavily used footpaths, are occupied by inferior birds or first-time breeders.

It is hard to imagine a more intrusive form of disturbance than the construction of a major road through woodland. Despite the fact that road improvement schemes have multiplied in recent years, we know rather little about their effects on birds and other wildlife. Evidence from studies in The Netherlands, however, indicates potentially serious implications for woodland birds. Densities of several bird species that breed in woodland adjacent to busy roads appear to be substantially reduced, compared with similar woodland further from the road (Reijnen & Thissen, 1987). These effects can be evident up to 500 m from the road. Furthermore, it has been demonstrated that woodland close to major roads is a relatively poor quality habitat for Willow Warblers (Foppen & Reijnen, 1994; Reijnen & Foppen, 1994). Male Willow Warblers nesting within 200 m of the roads comprised a higher proportion of yearlings, and were less likely to breed in the same area the following year, compared with birds nesting at some distance from the road. These results are important in showing that the impact of roads is not a simple matter of habitat loss. A serious reduction of habitat quality for many birds extends over a large area, probably as a consequence of the high noise levels generated by the traffic.

Creation of broadleaved woods on farmland

Tree planting is widely used in land restoration, and for landscape and amenity reasons. The most concerted efforts to create new woodland, however, have been on lowland farmland. In an attempt to reduce overproduction of farm crops, incentives have been available for planting trees on agricultural land since the mid-1980s. Initially, take-up was slow but the number of new woods appears to be steadily increasing, judging by the frequency with which patches of plastic tree shelters are appearing in many lowland landscapes. A high proportion of these woods are composed of native broadleaves as a result of government policy to restore the status of broadleaves in much of the lowlands. These new farm woods can have multiple purposes – to meet modest timber needs on the farm, to provide fuel, cover for game, and some income through timber sales.

There has been considerable interest in the potential for developing short-rotation forestry on farmland. Fast-growing hybrid poplars and willows can be coppiced on short rotations to provide mulch, pulpwood, chipwood, and fuelwood for heating and power generation. Large-scale planting of these crops has not yet materialised, though it may yet do so. It is too early to say how the bird communities in these new coppices might compare with those of traditional coppice. However, the birds breeding in two coppice plantations established on a peat-stripped bog in Ireland have been recorded by Kavanagh (1990). One of these plantations was composed of willow, the other of a mixture of broadleaves. These plantations held an interesting mixture of birds at quite a high density. Overall, the commonest species was Willow Warbler, which contributed an estimated 20% of the total number of territories, with a further 40% contributed by Reed Bunting, Chaffinch, Sedge Warbler and Wren. These Irish plantations are, however, not typical of those likely to be established on lowland British farmland. The survival of trees was poor at the Irish site, leading to a patchy canopy cover, which may have favoured a richer bird community than might otherwise have developed. The high numbers of Reed Buntings and Sedge Warblers probably reflected the wet nature of the site.

What are the implications for birds of creating new woods on farmland? The extent to which the creation of new woods will enrich the local bird life depends largely on the existing habitats. Even in the absence of woodland, many typical woodland birds manage to survive in farmed landscapes by living in the hedgerows. There is, however, much variation in the bird life of hedges, just as there is variation in the bird life of different woods. Hedges that are wide, contain mature trees, and are formed of many shrub species, tend to be richest in breeding birds (O'Connor & Shrubb, 1986). Even in landscapes with reasonable numbers of such hedgerows, several species of songbirds appear to be relatively scarce in hedges and are primarily dependent on woodland or scrub. These birds include Marsh Tit, Willow Tit, Treecreeper, Nuthatch and Nightingale. Apart from Whitethroat and Lesser Whitethroat, hedges are seldom an important habitat for warblers in Britain. On farmland, most territories of Willow Warblers, Chiffchaffs, Garden Warblers and Blackcaps are typically associated with patches of scrub and small woods, rather than with the hedges (Fuller *et al.*, 1995). There are exceptions of course. I know of one farm in south Norfolk where the rampant hedges surrounding its small fields are brimming with Willow Warblers in the spring and where Nightingales sing. But this place is a relic, an oasis of grass and scrubby hedges, in an otherwise stark arable landscape. It seems reasonable to conclude that the potential of new woods to increase the diversity of bird life is greatest in those districts where existing woods are few, and where hedges are of poor quality.

Afforestation of recently reclaimed Dutch polders (areas of land reclaimed from the sea) offers a striking illustration of how extensive tree planting can diversify the lowland avifauna. Since the 1940s, quite large forests of conifers and mixed broadleaves, including much poplar, have been planted on several polders. High densities and numbers of species have become rapidly established (R. Bijlsma, personal communication; Bremer, 1980; Verstrael, 1989). It took a comparatively short time for many of the plantations to acquire a more or less full complement of woodland bird species. After just 20–30 years of growth they often support large populations of species such as Woodcock, Turtle Dove, Hawfinch, Nuthatch and Marsh Tit. One reason for the rapid development of the woodland bird community may be that rates of tree growth are high. The poplars can reach 26–28 m after just 30 years. The relative isolation of the forests does not seem to have greatly impaired colonisation by woodland birds. The nearest potential source populations for many of the resident species were often more than 10 km distant. The Dutch experience shows that if the habitat is suitable, woodland birds will rapidly colonise new woodland. It is not always necessary to wait 100 years.

I shall now briefly discuss several aspects of the design and management of new woods in as much as they affect birds (see also Fuller *et al.*, 1995). Principles underlying the numbers of species, densities of birds, and the types of species found in woods were reviewed in Chapter 4 and these are relevant to predicting how birds will respond to new woodland.

The main design variables are the size, location and tree species composition of the wood. Though, individually, small woods support lower numbers of bird species than large woods, there are few species of British woodland birds that will not use small woods, say less than 3 ha, providing the habitat is suitable. Indeed, there is some debate as to whether a single large wood will hold as many species as several smaller woods of equivalent total area. Two papers have addressed this issue for British woods and both concluded that fewer species would be supported by one large wood (Ford, 1987; Woolhouse, 1987). An argument in favour of large woods is that their populations of birds and other species will be larger than those in small woods and, hence, will be less prone to extinctions. In practice, it seems desirable that landscapes should hold woods of a variety of sizes.

Should new woods be planted close to existing woods? For some woodland invertebrates and plants this is a vital consideration in maximising the chances of colonisation, but most birds can readily colonise quite isolated patches of woodland. There is, however, a potential benefit to be gained from allowing woodland to establish itself through natural regeneration adjacent to established woodland. By doing so, an interesting ecotone, grading from mature woodland to open farmland, could be created. Birds play a major role in dispersing the seeds of shrubs and trees (Fig. 9.3). Therefore, where natural regeneration is used to establish new woodland, it is possible that the provision of perches for birds may help to speed up the process (McClanahan & Wolfe, 1993). Where woods are planted, it is desirable to use a variety of trees native to the district and to plant underwood species as well as canopy species, especially close to the edges of the woods. This will help to develop structural and floristic complexity within the wood, both of which are important to birds (Chapter 4). Special treatments at the edges of woods offer one way of enhancing the value of small woods as wildlife habitats. For example, a narrow belt of scrub, cut piecemeal on a long rotation, could be maintained around the wood. This could benefit warblers and many insects too. Another possibility would be to carry out heavier thinning at the edge of the wood, with the aim of opening up the canopy and promoting the growth of a dense understorey.

There is much to learn about how birds actually live in landscapes

Fig. 9.3. Almost any patch of hawthorn scrub will provide a food supply for Redwings and Fieldfares, that frequently lasts until the end of December. Thrushes are largely responsible for dispersing the seed of hawthorns and this accounts for the rapidity with which the shrub will establish itself on land which is not grazed or cultivated. Birds are also responsible for dispersing the seeds of many other shrubs and trees. Where the establishment of woodland or scrub is sought by natural regeneration, the provision of artificial perches may help considerably in the initial establishment of shrubs and trees. Chris Rose.

that are a matrix of fields, hedges and woods. This is true both at the level of the individual and the population. Individual birds may make use of several habitats. Some hedgerow-nesting Chaffinches and tits may, for instance, collect much of their food within nearby woods. Alternatively, some woodland nesters may disperse widely after breeding and make substantial use of hedges. It is possible that hedges close to woods may, therefore, be used more heavily by some birds than hedges far away from woods. The idea that, for many birds, hedgerows are suboptimal habitats and woods are preferred habitats, has been around for a long time (Krebs, 1971; Murton & Westwood, 1974). Are hedgerow bird populations really maintained by surplus production from woodland, or by birds that are unable to establish territories in woodland? These questions have not been satisfactorily answered. The reality is unlikely to be simple because there is huge variation in the quality both of individual hedges and woods as bird habitats.

The case for natural woodland

Quite rightly, the emphasis of much habitat conservation in Britain is on management. Many scarce plants and animals depend on particular conditions that were created and sustained under traditional land-uses, but which have become obsolete during the twentieth century. Coppice and wood-pasture are prime examples. Woods which have a continuous history of coppicing or wood-pasture, and which are still managed in the traditional way, are rare indeed. Almost without exception, these places are refuges for rare species and it is vitally important that they continue to be treated in the traditional manner. It is also important that these treatments are restored in ancient woods where they have recently ceased. The case for management in many long-neglected ancient woods, however, is not so convincing. Nowhere is this more true than in old coppice. Conservation organisations now collectively own a huge area of ancient woodland in Britain and it has been suggested that an objective for some of these reserves should be the creation of natural woodland (Peterken, 1991; Fuller & Peterken, 1994). The concept of creating natural woodland areas has even been extended to upland conifer forests (Peterken *et al.*, 1992).

In North America, conservation efforts have long focused on protecting the still substantial areas of old-growth forest. During the 1980s, this grew into a major political issue with the emergence of the Northern Spotted Owl as a potent symbol of the magnificent forests of the Pacific north-west. These owls live mainly, but not exclusively, in old-growth characterised by large trees containing large cavities, much standing and fallen dead wood, and plenty of open space beneath and within a multilayered canopy (Thomas *et al.*, 1990). Over the last 100 years, this type of forest has been steadily reduced in total area and is becoming more fragmented. Against a background of intensifying controversy between conservation interests and the logging industry, a major attempt was made to establish a scientifically sound conservation strategy for the owl. The strategy was developed by a committee established by four United States agencies. The published strategy represents an impressive assessment of the status, trends, requirements and current risks to the owl (Thomas *et al.*, 1990). The strategy proposed the retention of large blocks of suitable habitat, each individually sufficient to meet the needs of several pairs of owls. It further suggested that these habitat blocks should be distributed throughout the range of the species, but that distances between them should not be so great as to prevent dispersing juvenile owls successfully emigrating from their natal block. Recommendations

Fig. 9.4. Old ash coppice in Suffolk which has long since been abandoned. It would not be easy to revive coppicing in a wood of this structure and a better solution might be to leave it unmanaged with the long-term aim of creating natural woodland. Peter Wakely, English Nature.

were also made that the landscape between the blocks should contain habitat that would sustain the birds during dispersal. The development of this strategy was notable also because it highlighted the severe difficulties that had to be faced in establishing a credible conservation plan for an endangered species when key pieces of ecological information were unavailable (Verner, 1992). Ultimately, the political issues raised by the old-growth forests and their Spotted Owls were sufficiently serious to merit the intervention of President Clinton, who sought to reach a compromise plan. This is not the place to analyse this conflict and its solution in greater detail, but it serves to illustrate the strength of feeling engendered by these untouched forests and their wildlife.

By contrast, in Britain there has been nothing remotely natural about the countryside for the last thousand years, so why should the idea of natural woodland be gradually gaining acceptance? There is no hope of recreating the wildwood but, given time, these woods will assume some characteristics of primeval forest (see Chapter 2 for a description of unmanaged forest in Poland). In Britain, we could expect such woods to contain much larger amounts of dead wood and much larger trees than we are accustomed to in managed woods (Kirby *et al.*, 1991). The natural woods would also be patchy, in an unplanned way, with small gaps created by the death of individual trees and larger ones by storms. Natural woods would provide unique sites for research and education, especially in relation to the natural dynamics of woodland. We cannot be sure what communities of plants and animals would develop in these woods, but this uncertainty is their essence.

As far as birds are concerned, there are no Spotted Owls waiting in the wings to exploit any expansion of old-growth in Britain. It would be startling if an increase in the extent of natural woodland led to radical changes to the British woodland avifauna. The value of these areas lies more in the baseline they would offer for measuring future impacts of human activities on woodland. The few comparisons that have been made of bird life in managed and unmanaged forests suggest that natural woodland might be richer in hole-nesting birds, and perhaps support higher overall densities of birds, than mature managed forest of similar tree species composition (Joensen, 1965; Nilsson, 1979c; Mannan & Meslow, 1984).

In reality, most future natural woodland in Britain will be small patches, typically less than 30 ha. While these woods will be extremely interesting, arguably the single most exciting development that could occur in British conservation would be a commitment to create a natural forest, sufficiently large to simulate the dynamics of the wildwood. Natural forests are in a state of constant flux, the main agents of change being fire, storms and disease. Fire is generally of greater significance in boreal forests than in the temperate zone. Major disturbances in natural forest are episodic. This was clearly shown by Foster (1988) who described the history of a 20 km² tract of virgin forest in New Hampshire. He drew a distinction between relatively frequent, localised disturbances (caused by individual treefalls, storms, lightning, fire and disease) and less frequent but more extensive events, which had profound effects on stand structure and development (Fig. 9.5). The latter were mainly hurricanes, 12 of which were recorded over a period of 350 years. The storm of October 1987, which devastated much woodland in south-east England, is a prime example of an occasional but highly significant disturbance. Our perception of the spatial structure and dynamics of forests depends strongly on the scale on which the system is viewed (Fig. 9.5). A unique opportunity to appreciate the patchiness of a natural forest in Europe is offered by Białowieża National Park in Poland which covers 47.5 km² (Falinski, 1986; also Chapter 2). As in the North American virgin forests, storms have intermittently created treefall gaps, sometimes extending over several hectares. Careful observation reveals a mosaic of patches varying greatly in size and at different stages of growth. Fallen dead trees are evident for decades after the windfall, long after trees have established themselves within the gap and the canopy has closed. Within sites of less than, say, 10 km² such patchiness would not be so evident.

Fig. 9.5. Damage caused by a hurricane in 1938 in the Pisgah Forest, New Hampshire. The black areas are lakes. The map demonstrates the patchy nature of damage caused by severe storms within natural forest. The small lines show locations of treefalls. From Foster (1988).

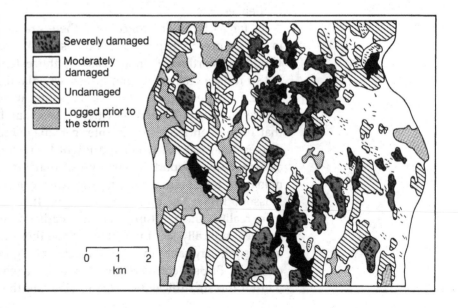

There is much to learn about the way that different species of birds respond to natural patchiness in forests such as that illustrated in Figs. 9.5 and 9.6. Far more is known about how birds respond to the patchiness created in managed forests as a result of felling. Earlier

Fig. 9.6. A treefall gap in Białowieża National Park, Poland. The fallen tree in the foreground is a huge Norway spruce. Note the large number of trees that remain standing within the treefall gap. Rob Fuller.

chapters have shown how bird communities undergo fairly predictable changes as the trees grow in conifer plantations, broadleaved high forest and in coppice. Similar changes in bird communities occur in natural forest but they are not so clearly defined as those in most managed forests. My own observations in Białowieża Forest show that the breeding bird communities in areas severely damaged by storms (large treefall gaps) are different from those in nearby undamaged forest. Species associated with treefall gaps include Tree Pipit, Chiffchaff, Garden Warbler, Blackcap and Dunnock. On the other hand, Red-breasted Flycatcher and Wood Warbler avoid storm-damaged areas and prefer a tall closed canopy. However, most 'mature forest species', including Nuthatches and woodpeckers, do make use of treefall gaps. The reason is that even the heavily damaged areas contain some standing live trees, as well as a huge amount of fallen trees and woody debris. This structure contrasts completely with that found in the young stages (the man-made gaps) of most managed forests. Therefore, the contrast between the bird communities of the early and late stages of forest growth is generally likely to be far greater in managed than in natural forests. Some species, however, probably prefer the types of gaps created by forestry. Red-backed Shrikes and Yellowhammers, for instance, are typical birds of young managed forests in Białowieża, but they rarely use treefall gaps in the National Park. One day it may be possible to make similar comparisons between managed and unmanaged forests in Britain.

Natural gaps can create new food resources and become focal centres of activity for some birds. Virkkala *et al.* (1991) described how Three-toed Woodpeckers benefited from the increase in numbers of dead and dying trees following a major storm in northern Finland. The devastation of elms by Dutch elm disease, both in woodland and in the wider British countryside, provides another example (Osborne, 1985). Woodpeckers and tits temporarily benefited from the large numbers of invertebrates living in the dead and dying wood, including the larvae of the beetles which actually spread the fungus. Other food resources that may be concentrated in treefall gaps are fruits and foliage insects. In North American woodland, migrating birds make heavy use of treefall gaps because they provide particularly rich feeding areas (Blake & Hoppes, 1986; Martin & Karr, 1986).

Conclusions

The reality of British woodland is that none exists in a natural state, and most is utterly unnatural. It seems entirely appropriate, therefore, that a large part of this book has focused on the effects of woodland

Fig. 9.7. It is entirely feasible to integrate good conservation management practices with productive forestry. This high forest oakwood in mid-Buckinghamshire is managed for timber production, mainly by small-scale felling. This will eventually create a patchwork of edges and different growth stages that will provide habitats for a wide range of plants and animals. By widening the rides, and cutting their margins on a piecemeal basis every few years, belts of bramble and shrubby regeneration have been created which offer new habitats for invertebrates and birds. Rob Fuller.

management on birds. The significance of woodland management on birds can be viewed on different scales. At the landscape scale, differences in bird populations from one district to another may be related to factors such as the total amount of woodland, and the degree to which it is fragmented. At the scale of the individual wood, the broad attributes of the bird community are likely to be influenced by the type of management system operating within the wood. At a finer scale still, the distribution of birds within a wood will be determined by vegetation structure and tree species composition. We need to understand how birds respond to their environments at each of these scales; it cannot be assumed that the answers at one scale can be transferred readily to another.

Different types of management are generally beneficial to different groups of plants and animals. The divergent character of the bird life associated with the basic woodland management systems in Britain has been described in this book. One goal of conservation within woodland nature reserves should be to ensure that not all woods are treated in the same way. Coppice, high forest and wood-pasture systems create complementary habitats and each of these has a place

in nature reserve management. We should also adopt policies of non-intervention in some woodland reserves so that our managed woods can be compared with more natural woodland in the future. In other reserves, there is also scope for exploring the conservation value of mixed woodland, shelterwood and selection silvicultural systems which are rarely used in Britain. If it can be demonstrated that these systems can successfully integrate timber production and conservation, it may be possible to encourage more landowners to adopt them. Principles for determining the objectives for individual woods are outlined by Fuller & Peterken (1994).

New opportunities have been opened up for nature conservation within British woodland. Destruction of ancient woodland, which was rampant in the post-war decades, has virtually ceased and a large number have been acquired as nature reserves. At the same time, there has been growing acceptance that conservation is a valid objective within commercially productive forestry. Against this background of success, it has become fashionable among the policy-makers of conservation to regard woodland as a low priority habitat. This is understandable given the critical threats facing many other habitats, such as old grassland and heathland. But wariness is needed to ensure that hard-won gains are not eroded. Recent reorganisation of the Forestry Commission has generated widespread concern about the future of state forests. One can only hope that the fine progress that has been made towards building sound conservation practices into the workings of these forests is maintained.

The key to long-term conservation success in woodland rests with a commitment to appropriate management. This applies within and outside nature reserves. Unfortunately, many management prescriptions are still not based on firm knowledge; they are often merely best guesses. At several places in this book I admit to having speculated about the effects of management on birds, so there is much work yet to be done. I am not an advocate of undertaking conservation management in woodland solely with birds in mind. There is a need to explore how much common ground exists between the needs of different groups of plants and animals, so that principles can be established with the widest benefits to wildlife.

Species of birds using woodland, forest and scrub habitats in Britain: their habitats, nest sites, food and feeding sites

Species have been arranged in four groups (A–D) according to the use they make of woodland and scrub during the breeding season, and are listed in taxonomic order (Voous, 1977) within each group. Unless stated to the contrary, habitat information refers to **breeding habitat in Britain**; where habitats are different elsewhere in Europe this is indicated. I have insufficient knowledge of Ireland to describe habitat usage there, but have indicated under *Status* if a species is absent from Ireland as a breeding bird. The following sources have been drawn on extensively and are not cited in the species accounts below: general information (Voous, 1960; Beven, 1964; Cramp & Simmons, 1977, 1979, 1982; Harrison, 1982; Cramp, 1985, 1988, 1992; Hutchinson, 1989; Cramp & Perrins, 1993; Gibbons *et al.*, 1993); tits (Perrins, 1979); crows (Goodwin, 1976); finches (Newton, 1972); nest sites (Campbell & Ferguson-Lees, 1972; Harrison, 1975).

Codes for British status: **R**, resident; **T**, tropical migrant; **S**, short-distance migrant wintering in southern Europe or North Africa; **(I)**, introduced species; **(M)**, 'marginal' or very small breeding population (less than approximately 300 pairs in Britain and Ireland combined, or recorded in fewer than 50 *BTO Atlas* squares (Gibbons *et al.*, 1993)); **WV**, winter visitor (following British Ornithologists' Union, 1992). All comments on *Status* refer to the breeding season, unless otherwise stated. Comments on major changes in breeding status in Britain since 1950 are based on Marchant *et al.* (1990) and Gibbons *et al.* (1993). Comments on distribution indicate some of the major discontinuities; for further information consult Gibbons *et al.* (1993).

Nest heights are given as the typical range, in metres, unless otherwise indicated. The term 'insects' here is used to include both insects and other small arthropods.

Part A. Species that will both breed and feed within closed-canopy scrub and woodland. Note that some of these species prefer dense scrub or thicket-stage woods rather than mature woods.

Honey Buzzard

Status T (M). Absent from Ireland.

Habitat and nest site Mainly broadleaved woodland but will use mixed or conifer stands. In The Netherlands, Bijlsma (1986) found nests in all three types of woodland, though most were in mixed forest. Will nest in quite open stands and small clumps of trees, though in Britain it is associated with heavily wooded areas and in The Netherlands it prefers forests larger than 250 ha (Bijlsma, 1993). The species nests in the canopy. Bijlsma (1993) reported a preference for Douglas fir and Norway spruce as nest sites.

Food and feeding site Wasp larvae, small mammals, birds, amphibians and reptiles. Digs for wasp larvae on the forest floor.

Goshawk

Status R (I, M). Increasing. The indigenous population became extinct in the nineteenth century but, since the mid-1960s, the species has become re-established through escaped birds and deliberate releases. Absent from Ireland.

Habitat and nest site Mature conifer plantations and old broadleaved stands. The nest is usually in a large conifer, typically three-quarters of the way up the trunk, at an average height of 18 m (Anon, 1989). In The Netherlands, birds prefer coniferous and mixed forest (Bijlsma, 1988).

Food and feeding site Feeds mainly on medium-sized birds. Unlike Sparrowhawk, the sexes show no habitat segregation for feeding and, in Sweden, the birds strongly prefer to hunt in large patches of mature forest, avoiding young successional stages (Widén, 1989).

Sparrowhawk

Status R; WV. Recent recovery from pesticide-induced sharp decline.

Habitat and nest site Wide range of thicket-stage and mature woodland, both broadleaved and coniferous. Highest-quality nesting places (those most regularly occupied and where breeding success is highest) are in late-thicket/early-pole stage; occupancy and success decline as trees exceed 17 m and become widely spaced (>5 m) (Newton, 1991). The nest is placed in the canopy or close to the upper trunk (5–15 m), often close to a ride or clearing.

Food and feeding site Mainly songbirds in woodland, farmland and gardens. The male hunts more within woodland than the female (Marquiss & Newton, 1982).

Capercaillie

Status R (I). Re-introduced to Scotland in the nineteenth century; now declining. Absent from Ireland.

Habitat and nest site Conifer woods in the eastern Highlands, both semi-natural ancient forests and plantations. The most productive breeding habitats appear to be mature stands of Scots pine with well-spaced trees and a vigorous field layer of bilberry. The density of cocks at most Scottish leks is higher at sites typified by a native pine forest structure with old 'granny' trees and a good field layer (Picozzi *et al.*, 1992; Moss & Picozzi, 1994). In Scandinavia,

the species is strongly associated with old conifer forests, the availability of which is the main factor determining distribution of lekking sites (Rolstad & Wegge, 1987). In winter, adult males remain in the old forest close to the lek, but females and subadult males may range more widely making use both of old and middle-aged stands (Gjerde & Wegge, 1989). Nests on the ground.
Food and feeding site Main winter food is needles and buds of Scots pine taken from the canopy. Also feeds on heather and bilberry on the ground. Chicks eat many invertebrates.

Pheasant
Status R (I). Its distribution is affected greatly by hand-rearing.
Habitat and nest site Wide range of woodland types. Also occurs widely outside woodland. Its preferred woods, both in winter and for breeding, are ones with much farmland edge and with dense shrubby interiors (Robertson, 1992; Robertson *et al.*, 1993). Males establish territory along wood edge, but females make more use of woodland interior. Nests on the ground.
Food and feeding site Plant and animal food, shoots, seeds and berries. Chicks eat many invertebrates. Feeds on the ground.

Golden Pheasant
Status R (I, M). Absent from Ireland.
Habitat and nest site Local bird of dense pine plantations and southern yew woods. Also deciduous or mixed woods with a dense understorey of rhododendron. In Thetford Forest it uses 15 to 30 year-old thicket-stage plantations (Bowden & Hoblyn, 1990). Nests on the ground.
Food and feeding site Presumed to be mainly herbivorous. Feeds on the ground.

Lady Amherst's Pheasant
Status R (I, M). Apparently declining. Absent from Ireland.
Habitat and nest site Local bird of dense pine plantations and deciduous or mixed woods with thick undergrowth, often rhododendron. Nests on the ground.
Food and feeding site Presumed to be mainly herbivorous but known to eat some invertebrates. Feeds on the ground.

Woodcock
Status R; WV. Probably in decline. Patchily distributed in Ireland.
Habitat and nest site Broadleaved, coniferous and mixed woods with a substantial field layer. One study (Hirons & Johnson, 1987) found that nests were placed in bramble areas. In Thetford Forest it uses thicket-stage plantations (Bowden & Hoblyn, 1990). Nests on the ground.
Food and feeding site Both in summer and winter, earthworms are a major food (Hirons, 1982; Granval & Muys, 1992). Hirons & Johnson (1987) recorded a preference for feeding in areas with a soil of high pH, a ground cover of dog's mercury and avoidance of beech. In winter, it will feed nocturnally outside woodland on pasture. Always feeds on the ground.

Woodpigeon
Status R; WV.
Habitat and nest site Breeds in a wide range of woodland and scrub. Also breeds widely outside woodland, in hedgerows, parkland and large gardens. In conifer plantations, it reaches highest numbers in the thicket stage. In eastern Europe, it is largely confined to extensive forest. Most nests are in bushes or in the low to mid-canopy (2–10 m) though nests have been recorded as high as 25 m.
Food and feeding site Herbivorous. A range of seeds, fruits, buds, leaves and some invertebrates; exploits foods from both woodland and cultivated land. Feeds on the ground, in the tree canopy and in the shrub layer.

Cuckoo
Status T.
Habitat and nest site A nest parasite. Uses many habitats, including woodland and scrub where the main hosts are Robin and Dunnock (Glue & Morgan, 1972).
Food and feeding site Insectivorous, mainly caterpillars. Feeds mostly on the ground.

Tawny Owl
Status R. Absent from Ireland.
Habitat and nest site All types of woodland though avoids young stages of growth. Widely distributed outside woodland, on farmland and in urban areas wherever there are suitable nest sites. Nests mainly in tree-holes, also on the ground, in buildings, and in old nests or dreys. Takes to special nest boxes.
Food and feeding site Mainly rodents, but also birds, amphibians and worms. Mainly a 'sit and wait' hunter, catching most food on the ground, though it will also cruise around bird roosts.

Wryneck
Status T and S (M). Decline almost to extinction but still breeds in most years in England and Scotland. Absent from Ireland.
Habitat and nest site Previously widespread in mature woodland and parkland, though most woodland birds probably occurred at the edge. The Scottish birds breed in open deciduous woodland dominated by birch. Nests in tree-holes and will use nest boxes.
Food and feeding site Mainly ants which are taken on the ground and from tree trunks and branches. Does much feeding outside woodland in grassland.

Green Woodpecker
Status R. Absent from Ireland. More abundant in southern than northern Britain.
Habitat and nest site Mainly broadleaved and mixed woodland, parkland and farmland with many mature trees. Will use mature conifer plantations, perhaps mainly in the north of its range. Tree-hole nester: excavates its own hole, typically in a live tree (2–15 m).
Food and feeding site Invertebrates including many ants. Does much feeding outside woodland in grassland. Much foraging on the ground but also on the trunk and large limbs of trees.

Great Spotted Woodpecker

Status R; WV. The northern sub-species is a passage and winter visitor to Britain, especially in irruption years. Absent from Ireland.

Habitat and nest site Mature woodland, both broadleaved and coniferous. Also breeds in well-timbered farmland and parks. In southern Europe, it is mainly found in mountain forest. Excavates its own nest hole, typically in a live tree (2–14 m).

Food and feeding site Invertebrates. Seeds in winter. In winter, takes many dead wood insects both from standing and fallen timber. Winter food in boreal forests consists of pine and spruce seeds. In summer, feeds heavily on defoliating caterpillars (Smith, 1987).

Lesser Spotted Woodpecker

Status R. In decline after expansion during the 1970s which may have been caused by Dutch elm disease. Absent from Scotland and Ireland.

Habitat and nest site Mature woodland, mainly broadleaved. Also found in well-timbered parkland and farmland. In Sweden, prefers old broadleaved stands containing many standing dead trees (Olsson *et al.*, 1992). Excavates its own nest hole in trees at heights from 1–15 m. Shows stronger selection of dead trees for nesting than does Great Spotted Woodpecker (Hågvar *et al.*, 1990).

Food and feeding site Mainly insects. Often feeds on small branches high in the canopy unlike other woodpeckers.

Wren

Status R; WV. Fluctuates widely in numbers depending on winter weather.

Habitat and nest site Wide range of scrub and woodland habitats as well as gardens, hedgerows etc. In woodland, the highest densities are found where there is dense low cover, such as bramble. Often reaches higher densities at the edge of the wood than in the interior. In eastern Europe, however, Wrens are mainly confined to large blocks of forest. The nest is typically situated in dense, low cover or concealed against a tree trunk in honeysuckle etc. The nest is usually within 2 m of the ground.

Food and feeding site Mainly insects. Feeds close to the ground, in the field layer or in the low shrub layer.

Dunnock

Status R; WV.

Habitat and nest site Scrub, young coppice and plantations, gardens, hedges etc. In mature woodland, it is often confined to the edge. In central, northern and eastern Europe it is found mainly in extensive coniferous or mixed forest. Generally needs a dense low cover. The nest is well concealed in a bush or in the field layer (0.1–2 m).

Food and feeding site Mainly insects. Many seeds in the winter. Feeds mainly on the ground, also in the field layer and the low shrub layer.

Robin

Status R; WV.

Habitat and nest site Woodland, scrub, gardens, hedgerows and other places with bushes and trees. Avoids very young woodland and scrub. In a Sussex wood, Robins chose patches with dense vegetation within 3 m of the ground and their breeding success was greatest in such areas (Hoelzel, 1989). In Belgium, wintering Robins also prefer habitats with a dense shrub layer (Adriaensen & Dhondt, 1990a) and birds breeding in woodland are more migratory than those in gardens and parks (Adriaensen & Dhondt, 1990b). In central and eastern Europe, it occurs mainly in large areas of forest; in southern Europe it is mainly a bird of montane and sub-alpine forest. Nests on or close to the ground, often in walls, sometimes in tree-holes. In Polish primeval forest, tree cavities are commonly used.

Food and feeding site Mainly insects and other small invertebrates, also fruit and some seeds outside the breeding season. Feeds mainly, but not exclusively, on the ground.

Nightingale

Status T. Long-term contraction of range. Restricted to central and south-east England. Absent from Ireland.

Habitat and nest site Woodland and scrub with a dense understorey. Appears to choose habitats with a combination of patches of bare or sparsely vegetated ground, yet with dense foliage below 2 m. Such conditions are provided by some coppice just after canopy-closure, often 6 to 8 years after cutting. Will nest in young conifer plantations with many broadleaved, non-crop trees. Nests on or close to the ground, often concealed in bramble or dense field layer foliage.

Food and feeding site Mostly small ground-living invertebrates and berries late in the summer. Feeds on or close to the ground.

Redstart

Status T. Recovering from a 1970s decline. Virtually absent from Ireland. Scarce in south and east Britain.

Habitat and nest site Main habitat in Britain is mature, often rather open, woodland. It is now most characteristic of upland sessile oak and birch, also of New Forest wood-pasture and of northern pine woods. Also breeds on lowland heaths and commonland, especially where there are scattered trees, and in hedges with old pollards. Mainly nests in tree-holes and crevices, or in a hollow or wall. Nests are usually lower than 8 m. Uses nest boxes.

Food and feeding site Predominantly insects. Feeds mainly on the ground and in the canopy, but also 'flycatches'.

Blackbird

Status R; WV.

Habitat and nest site Widely distributed in all habitats with trees or bushes. In many woods it reaches highest breeding densities at the edge. In eastern Europe, it occurs mainly in extensive forest tracts. Usually nests in the shrub layer, though often uses artifacts, and sometimes tree-holes or even on the ground. Tree-holes are a typical nest site in Polish primeval forest. Nests are seldom placed higher than 4 m.

Food and feeding site Insects and worms taken on the ground. Eats berries of many species of shrubs. Often feeds in the open, some distance from the nest.

Song Thrush

Status R; WV. Decline since the 1970s.

Habitat and nest site Similar to Blackbird. In eastern Europe, it is found mainly in extensive tracts of forest.

Food and feeding site Wide range of invertebrates including many snails. Otherwise similar to Blackbird, though more reluctant to feed away from cover than other large thrushes.

Redwing

Status R or S (M); mainly WV. Scarce breeder, virtually confined to Scotland. Absent from Ireland. Widespread and abundant in winter.

Habitat and nest site Breeds in upland birchwoods, woods in the grounds of large houses, etc. In Europe, it breeds in a wide variety of scrub and woodland, from central Europe to the far northern taiga. The nest is usually placed in the fork of a bush or tree (0.25–7 m).

Food and feeding site Invertebrates and berries. Winter flocks feed in the canopy of berry-bearing bushes or on the ground, both in woodland and in open fields.

Garden Warbler

Status T. Rarer in Ireland than in Britain, where it is more abundant in the south than in the north.

Habitat and nest site Mainly broadleaved woodland and thick scrub with very dense foliage in the lower 2–3 m. In many mature woods, it is confined to the external edge. High densities occur in downland scrub and in coppice around the canopy-closure period. In southern Europe, it occurs mainly in mountain forests. The nest is usually concealed inside a bush, or in dense field layer vegetation, within 2 m of the ground, but not on the ground.

Food and feeding site Insects. Berries in late summer. Feeds in the shrub layer and in low canopy foliage.

Blackcap

Status T and S; WV. Increasing. Patchily distributed in Ireland. Scarcer in northern than southern Britain.

Habitat and nest site Similar to Garden Warbler with which it can compete and exclude from potentially suitable habitat (Garcia, 1983). Will use taller, older scrub than Garden Warbler and is more frequent within mature woodland with a well developed shrub layer. Nests in tall bramble, a bush or in a low tree, usually no higher than 3 m.

Food and feeding site Insects and berries. Feeds in the shrub layer and in low canopy foliage.

Wood Warbler

Status T. Very rare in Ireland. Within Britain, relatively rare in the south and east.

Habitat and nest site Mature broadleaved woodland with a sparse shrub layer (Bibby, 1989). Major habitats are western and northern oakwoods, birchwoods and New Forest woodland. Has been found breeding in 100 year-old Douglas fir in Wales (Currie & Bamford, 1982) though nests more frequently in coniferous forest in eastern Europe. In some woods, territories are clustered, and locations of clusters can change from year to year (Herremans, 1993). Nests on the ground.

Food and feeding site Insects, which it catches mainly in the canopy (Stowe, 1987).

Chiffchaff

Status T and S; WV. Relatively scarce in northern Britain.

Habitat and nest site Various woodland types ranging from mature broadleaved to middle-aged coppice. Appears to need both tall trees and some open spaces (e.g. rides, clearings, sparse canopy). Nests close to ground in dense grass, bramble, etc.

Food and feeding site Insects. Feeds in both the shrub layer and the canopy. Saether (1983) found that it fed mainly in the tree layer, slightly higher than the Willow Warbler.

Willow Warbler

Status T.

Habitat and nest site A bird of scrub, young woodland growth and woodland edges. Generally reaches peak densities in scrub, coppice and young plantations, both broadleaved and coniferous, just before and during canopy-closure. Unlike Chiffchaff, it is usually confined to the extreme edge of mature woodland. Nests on, or very close to, the ground in dense cover.

Food and feeding site Insects. Feeds in shrub layer foliage and in the canopy. Saether (1983) found that it fed mainly in the tree layer, slightly lower than the Chiffchaff.

Goldcrest

Status R; WV. Fluctuates widely depending on winter weather.

Habitat and nest site Though highest densities occur in mid-stage and mature coniferous woods, it breeds also in many purely broadleaved woods and sparingly in dense scrub. Over much of Europe, however, it is largely restricted to coniferous woodland. It nests in the coniferous canopy. In broadleaved sites, nests are often supported in ivy or honeysuckle in the low canopy. Nest heights range from 2–12 m.

Food and feeding site Insects. Feeds mainly in the tree canopy and in the shrub layer. In broadleaved woods, in winter, it may feed mainly in the shrub layer (Beven, 1959). Gibb (1954) recorded that it fed more on twigs and buds than on leaves.

Firecrest

Status R or S (M); WV. Absent from Ireland.

Habitat and nest site Mature coniferous or mixed woods at least 8 m tall, especially Norway spruce or Douglas fir, occasionally in broadleaved stands with holly (Batten, 1973). More commonly uses broadleaved woodland in Europe. Its nest site is similar to Goldcrest (2–14 m). In winter, in south-west Britain, it occurs in coastal scrub and in other habitats with dense cover near water.

Food and feeding site As for Goldcrest, but takes larger prey and feeds more on larger branches (Leisler & Thaler, 1982).

Spotted Flycatcher

Status T. In long-term decline.

Habitat and nest site Mature woodland, mainly broadleaved (though it will use mature conifers), parkland and large gardens. In woodland, it prefers stands with an open canopy or open spaces including treefall gaps, rides and the edges of felling coupes. Its nest site is variable: in an open cavity or crevice, on top of a wide branch, against a trunk supported by a small branch, in a creeper. Will use an open nest box. Nest height up to 10 m.

Food and feeding site Insects, caught mainly in flight.

Pied Flycatcher

Status T. Virtually absent from south-east Britain and from Ireland.

Habitat and nest site Mature broadleaved and coniferous woodland. Highest densities are in broadleaved woodland, typically western oakwoods. Even when a superabundance of nest boxes is provided, densities are higher in deciduous than coniferous woodland (Lundberg & Alatalo, 1992). Preferred woods generally have a sparse understorey. Nests up to 15 m.

Food and feeding site Insects. Young are fed many caterpillars. Uses varied feeding sites. In Welsh woods, Stowe (1987) found that birds fed most commonly in the canopy and on the ground.

Long-tailed Tit

Status R. Population fluctuates in response to cold winters.

Habitat and nest site Scrub, woodland and hedges. Scarce breeders in coniferous woods. In many woods, it is commonest at the edges where the vegetation is often bushier. Nests in dense, often thorny, thickets and hedges, a few metres above the ground, or, occasionally, much higher in the fork of a tree or against the trunk, up to 20 m.

Food and feeding site Insects and small seeds, though eats fewer and smaller seeds than the *Parus* tits. Feeds both in the shrub layer and in canopy trees where it feeds in the outer twigs. Feeds more from twigs and buds than *Parus* tits (Gibb, 1954).

Marsh Tit

Status R. Its range barely extends into Scotland and it is absent from Ireland.

Habitat and nest site Broadleaved woodland, avoiding sites with very little understorey. Sometimes in scrub and gardens. Nests in a hole in a live tree, often low down, typically within 3 m of the ground. Rarely uses nest boxes.

Food and feeding site Insects in the breeding season, seeds and fruits at other times, including beechmast. Feeds in the low canopy, in the shrub layer and on the ground. In summer, it may prefer to forage in the shrub layer rather than in the canopy (Beven, 1959). Commonly stores food.

Willow Tit

Status R. Virtually absent from northern Scotland. Absent from Ireland.

Habitat and nest site Broadleaved and coniferous woodland, though far less typical of the latter than elsewhere in Europe. Reported as characteristic of pine in Thetford Forest (Gibb, 1960) but now it is absent from, or very rare in, these stands. Broadleaved sites are often marshy, sometimes scrubby. In Scandinavia, Willow and Marsh Tits are far more distinctly separated by habitat, the former being typical of conifer forests, the latter of rich deciduous woodland (Alatalo & Lundberg, 1983). Excavates its nest hole low in a rotten stump or tree, within 2 m of the ground. Does not use conventional nest boxes.

Food and feeding site   Insects, with small seeds in winter. Feeds in low foliage and on dead wood, rarely on the ground. Stores food.

Crested Tit

Status R. Confined to north-east Scotland. Absent from Ireland.

Habitat and nest site Scots pine woodland, both plantations and ancient forests. Does not use plantations less than 20 years old (Cook, 1982). Excavates its nest hole in a stump or dead tree, mostly within 3 m of the ground. Occasionally uses nest boxes.

Food and feeding site Insects, with many seeds and berries outside the breeding season. Feeds mainly in Scots pine, both in the tree canopy and in low bushes, and also on the ground in heather (Hartley, 1987). Will store both insect food and seeds.

Coal Tit

Status R.

Habitat and nest site Highest densities are in coniferous woodland, especially late-thicket and mature stages. Common in a wide range of broadleaved and mixed woods. One of the most abundant breeding birds in derelict Scottish oak coppice (Williamson, 1976). Over much of Europe, however, Coal Tits are strongly confined to conifer forest. Hole-nester, frequently using holes at ground level in tree roots and stumps, as well as tree-holes up to 5 m. Readily uses nest boxes.

Food and feeding site Mainly insects in summer and a mixture of insects and seeds in winter. Feeds mainly in the canopy though makes use of branches and trunks of broadleaved trees in winter (Beven, 1959). Regularly stores food.

Blue Tit

Status R; WV.

Habitat and nest site Essentially a bird of broadleaved woodland. Will breed at low density in coniferous woodland, though not so frequently as Great Tit. Common in gardens, parkland and farmland with mature trees. Nests in tree-holes up to 15 m. Readily takes to nest boxes.

Food and feeding site Insects, with seeds in winter. Feeds mainly in the canopy and in the shrub layer.

Great Tit

Status R.

Habitat and nest site Primarily a bird of broadleaved woodland though occurs in a wide range of woods and scrub, also gardens, parkland and hedges with trees. Nests in a hole in a tree, a wall or in rocks. Tree nests are located at heights up to 15 m. Takes to nest boxes more readily than any other tit.

Food and feeding site Insects, with many seeds in winter. Eats larger seeds than other tits, including beechmast. Feeds low down in the shrub layer, on low trunks and on the ground.

Nuthatch

Status   R. Its range barely extends into Scotland and it is absent from Ireland.
Habitat and nest site   Mature broadleaved woodland and parks and gardens with large trees. In Sweden, densities are similar in oak and beech woods but breeding success is highest in oak; densities are relatively low in spruce forest though breeding success is high there in years when spruce seeds are abundant (Nilsson, 1976). In Belgium, the 'best' territories contain oaks but not conifers, and birds prefer to forage on oak than beech (Matthysen, 1990). In contrast, a German study found no clear difference in breeding success between tree species (Schmidt *et al.*, 1992). Conifers are rarely used in Britain. Nests in holes in live trees up to 20 m but will use nest boxes.
Food and feeding site   Insects in summer, together with seeds in winter. Eats large seeds including beech and hazel nuts. Population densities can correlate closely with the size of the beech or hazel crops (Nilsson, 1987; Enoksson, 1990). Gleans insects from the bark of trunks, large branches and fallen trees, both live and dead. Takes seeds on the ground.

Treecreeper

Status   R.
Habitat and nest site   Woodland, both broadleaved and coniferous though the former probably carries greatest densities. In mainland western Europe, it breeds mainly in coniferous forest (though Short-toed Treecreeper uses lowland broadleaved forest). Avoids young-growth. Also breeds in well-timbered parks and gardens. Nests behind loose bark or in crevices, as low as 1 m. Will use special nest boxes.
Food and feeding site   Insects which are found by gleaning the bark of trunks and branches. Makes less use of branches than Nuthatch (Beven, 1959). Forages mainly on living trees though makes more use of dead trees in winter (Gibb, 1954).

Golden Oriole

Status   T (M). Absent from Ireland.
Habitat and nest site   Broadleaved woodland. In Britain, nearly all breeding birds are thought to be in mature poplar plantations. Nest placed in canopy.
Food and feeding site   Large insects and fruits. Feeds mainly in the canopy.

Jay

Status   R; WV. Occasional Continental irruptions. Far more abundant in the south than in the north of Britain. Patchily distributed in Ireland.
Habitat and nest site   Though it now uses parks and gardens, even urban areas, it remains essentially a woodland bird, using both broadleaved and coniferous woods. Thicket-stage plantations and woods with a fairly dense shrub layer are suitable. Its nest is usually placed 2–6 m above the ground in the crown or fork of a small bush or tree, or in a creeper.
Food and feeding site   Has a mixed diet, including birds and eggs. Acorns are eaten for much of the year; birds store many acorns in autumn (Bossema, 1979). Beech seed is a relatively minor food, though it may be stored when acorns are scarce, and in some regions the Jay is the major dispersal agent for both oak and beech (Nilsson, 1985). Nestlings are fed caterpillars as well as acorns. The feeding site is variable and includes both the ground and foliage.

Chaffinch

Status   R; WV.

Habitat and nest site   Wide range of woodland types both broadleaved and coniferous; also gardens, parks and hedges. The most abundant breeding bird in many broadleaved woods. Highest densities are in broadleaved woods; spruce generally holds higher densities than pine. It generally avoids young woodland, becoming commoner after canopy-closure, though young coppice with many standards can support many birds. In winter, it feeds mainly on farmland, except in years with plentiful beechmast. Roosts communally in evergreens, bramble and conifer plantations. The nest is placed either in the canopy or in a bush, typically in a fork. The range of nest heights is wide (1–20 m).

Food and feeding site   Mainly insects in summer; switches to seeds in winter, including beechmast in good mast years. Insects are taken mainly from the shrub layer and the canopy; seeds are eaten on the ground.

Greenfinch

Status   R; WV.

Habitat and nest site   As a woodland bird, it is mainly associated with young conifer plantations, young and mid-growth coppice, and the edges of more mature woodland. It also breeds in mature gardens, thick scrub, churchyards, dense hedges etc. Though birds can be found in woods in winter, the main feeding habitats at this time of year are farmland, gardens and saltmarshes. Birds roost communally, mainly in woodland or dense scrub, preferring sites with bramble or evergreen e.g. rhododendron. The nest site is in a bush or creeper (1–5 m).

Food and feeding site   Seeds of woodland, farmland and saltmarsh plants. A wide range of seed sizes, but larger ones are selected. In winter, many birds now depend on garden feeding. Sitka spruce seeds have been eaten increasingly in recent years (Shaw & Livingstone, 1991). Feeds both on the ground and on the plants themselves.

Siskin

Status   R; WV. Major increase with spread of coniferous forestry but remains more common in the north than south of Britain.

Habitat and nest site   Formerly a bird of old pine forests in north-east Scotland but now widely distributed as a breeding bird in mature pine and spruce plantations. Has adapted well to Sitka spruce. In winter, it occurs not only in conifer woods but also in birch and alder. Roosts communally in conifers and scrub. In winter it uses garden bird feeding stations. It nests in the conifer canopy, up to 20 m.

Food and feeding site   Conifer seeds and, in winter, alder and birch seeds. Most tree seeds are taken in the canopy but also feeds on the ground beneath birch and alder. Compositae seeds are taken when tree seeds are scarce. In Finland, birds are dependent on spruce seed in early spring, even when nesting in pine, and the density of breeding birds fluctuates in parallel with the size of the spruce cone crop (Haapanen, 1966).

Redpoll

Status R; WV. Former increase, now decreasing.

Habitat and nest site Two main habitats: (1) birch scrub or thickets of alder, thorn and willow; (2) young conifer plantations. In winter, it uses woods and places with abundant small seeds. Roost sites are similar to breeding habitats. It nests in a bush, 1–5 m above the ground.

Food and feeding site Seeds. Adults eat some insects in the spring. Birch seed is a major food, also alder seed in winter. Many other small seeds are eaten. Depending on the food being taken, the bird will feed in trees, in the field layer and on the ground.

Common Crossbill

Status R; WV. Movements, including irruptions, are extremely complex and are driven by food availability. The recent increase in numbers and range has almost certainly been caused by the expansion of coniferous forestry. Crossbills have colonised Ireland in the last 20 years or so.

Habitat and nest site Mature conifer forest, mainly spruce and pine plantations. Nests in the canopy, at heights from 8 m to more than 20 m.

Food and feeding site Conifer seeds: pine, spruce, larch and fir. Typically feeds high in the canopy.

Scottish Crossbill

Status R. Confined to north-east Scotland. Absent from Ireland.

Habitat and nest site Mature, semi-natural Scots pine forest. Its nest site is as for Common Crossbill.

Food and feeding site Pine seeds which are taken from cones in the tree.

Parrot Crossbill

Status (M). An occasional breeder in Britain; essentially nomadic. Not recorded breeding in Ireland.

Habitat and nest site Uses mature pine forest. Nest site same as other crossbills.

Food and feeding site As for Scottish Crossbill.

Bullfinch

Status R.

Habitat and nest site Scrub, hedgerows and woodland. In mature woods, it is often found mainly at the edge where the undergrowth is thickest. Mainly in broadleaved woods but also in thicket-stage conifers. In central and eastern Europe, however, it lives mainly in coniferous forest. Roosts in thorn bushes and evergreens. The nest is placed in a thick bush, 1–3 m above the ground.

Food and feeding site Seeds, buds, tree flowers and berries. Eats more buds and flowers than other finches. Nestlings are fed a mixture of seeds and invertebrates. Bullfinches feed mainly in the shrub layer.

Hawfinch

Status R. Absent from Ireland.

Habitat and nest site Mature broadleaved woodland, occasionally orchards and parks. Will use evergreens for roosting, also thorn bushes. The nest is placed on a branch, typically in the low or mid canopy (2–10 m).

Food and feeding site Mainly large fruits of woodland trees, especially hornbeam. Eats many insects in spring and the young are fed a mixture of insects and seeds. Takes seeds and fruit, both on the ground and in the trees.

PART B. Species that, when using scrub or woodland, are confined to habitats with an open canopy and small trees or bushes (e.g. very young plantations, young coppice and ecotones).

Hen Harrier
Status   R; WV. Patchily distributed in north-west Britain.
Habitat and nest site   Breeds in establishment-stage conifer plantations, up to 15 years, though rarely in restocked sites (Petty & Anderson, 1986; Bibby & Etheridge, 1993). Moorland is the other main habitat. A small fraction of winter roosts is in plantations (Clarke & Watson, 1990). Nests on the ground, though has nested successfully in the canopy of thicket-stage Sitka spruce in Ireland (Scott *et al.*, 1991).
Food and feeding site   Birds and small mammals, but mainly birds in the breeding season. Hunts over open ground and young conifers.

Black Grouse
Status   R. Widespread decline since the 1960s. Absent from Ireland. Confined to northern upland Britain and to north Wales.
Habitat and nest site   The Black Grouse lives at the interface between woodland and moorland. It uses a mosaic of habitats with different feeding and vegetation requirements at different seasons and stages of breeding (Parr & Watson, 1988; Cayford & Hope Jones, 1989; Baines, 1990; Cayford, 1993). Both broadleaves (especially birch) and conifers (both native pine and plantations) are used, though dense stands are avoided. In some regions (e.g. north-east England), the species occurs in sheepwalk areas with very little woodland or heather. Nests on the ground.
Food and feeding site   Mainly plants: buds, catkins, needles, shoots and berries. Feeds both on the ground and in trees. Young chicks eat many invertebrates.

Short-eared Owl
Status   R; WV. Has bred sporadically in Ireland. Patchily distributed in upland Britain.
Habitat and nest site   Nests on the ground among young conifers or in moorland. Will use restocks within existing forest, as well as afforestation. Occasionally breeds in sand dunes and saltmarshes.
Food and feeding site   Similar to Hen Harrier, though more dependent on small mammals.

Nightjar
Status   T. Decline since the 1960s. Very rare in Ireland.
Habitat and nest site   Young forestry plantations and heathland are the main habitats. Some birds nest in young sweet chestnut coppice. Plantations are used up to some 12 years of age (Bowden & Hoblyn, 1990). On heathland, Nightjars often nest in areas of scattered birches or pine, and such sites may even be preferred. However, if the trees are allowed to form a closed stand the habitat will cease to be suitable. The bird is a ground-nester.
Food and feeding site   Large insects which are taken in flight. In Dorset,

Alexander & Cresswell (1990) found that birds moved several kilometres from their nesting areas at night, presumably to feed, when they selected deciduous woodland and avoided coniferous plantations.

Woodlark
Status R and S?. Decline since the 1950s. Absent from Ireland.
Habitat and nest site There are two main concentrations in Britain: in the south (mainly Surrey/Hants/Dorset) on heaths and other open habitats; and in East Anglia, mainly in coniferous plantations up to some 5 years of age (Bowden, 1990). The birds seem to require trees or other tall features as lookouts or songposts. Nests on the ground.
Food and feeding site Mostly insects in the breeding season, mostly seeds in winter. Feeds on the ground, preferring areas with bare ground and short vegetation (Bowden, 1990).

Tree Pipit
Status T. Gradually declining. Virtually absent from Ireland.
Habitat and nest site In south and east Britain, uses mainly open habitats, e.g. young coppice and plantations, scrubby commons, heaths and downland. In the north and west, it is typical of upland sessile oakwoods and birchwoods. Also occurs in open Scottish pinewoods. Seems to need at least a few trees as songposts. Nests on the ground.
Food and feeding site Mostly insects. Feeds mainly on the ground in sparsely vegetated areas.

Whinchat
Status T. Decline since the 1950s. Much scarcer in Ireland than in Britain where it is most abundant in the west and north.
Habitat and nest site High densities are found in young upland conifer plantations with trees up to 3 m in height (Sykes *et al.*, 1989). Also breeds on rough grassland (e.g. Welsh *ffridd*), low moors and heaths with scattered bushes. Nests on, or close to, the ground.
Food and feeding site Small invertebrates and some seeds. Hunts from a perch, taking food from the ground, low foliage, or in the air.

Stonechat
Status R and S. Major decline since the 1970s. In Britain, it shows a marked westerly bias in its distribution.
Habitat and nest site Will nest in young conifer plantations but its main habitats are lowland and coastal heath and moors, especially with gorse and heather. Also in sand dunes and coastal scrub. More of a lowland bird than Whinchat. Nests on or close to the ground.
Food and feeding site Small invertebrates. Feeding site similar to Whinchat.

Grasshopper Warbler
Status T. Has decreased strongly since the 1960s.
Habitat and nest site Fens and various types of scrubby habitat where the canopy is very open and the field layer prolific. Young open conifer plantations are another major habitat. Nests on or close to the ground in dense herbage.
Food and feeding site Insects. Feeds on, or close to, the ground in dense cover.

Dartford Warbler

Status R (M). Fluctuates widely depending on winter weather. Confined to southern England. Absent from Ireland.

Habitat and nest site Lowland heaths with a mixture of heather and gorse. The most productive territories are ones with much gorse while the preferred nest site is in heather and is typically placed within 1 m of the ground (Bibby, 1979b). In southern Europe, it breeds mainly in maquis and dense thorny scrub.

Food and feeding site Insects. Feeds mainly in gorse (Bibby, 1979a).

Lesser Whitethroat

Status T. Scarce in much of northern and western Britain. Virtually absent from Ireland.

Habitat and nest site Tall scrub, though avoids areas with an extensively closed canopy. Another main habitat is tall dense hedgerows. In Europe, it commonly breeds in urban parks. Rarely uses coppice or young plantations though occasionally breeds on the woodland edge. The nest is usually placed in a bush within 2 m of the ground.

Food and feeding site Mostly insects, which are taken from the foliage of bushes.

Whitethroat

Status T. Lower numbers since 70% decrease between 1968 and 1969.

Habitat and nest site All sorts of low, open scrub including very young coppice, young plantations and low hedges. Avoids closed-canopy scrub. Nests in low dense cover, within 2 m of the ground.

Food and feeding site Mainly insects; small fruits in late summer. Feeds in the field layer and in low bushes.

Red-backed Shrike

Status T (M). Decline almost to extinction. Absent from Ireland.

Habitat and nest site Open scrub or heath with tall bushes. Typically nests in a thorny bush within 2 m of the ground.

Food and feeding site Mainly large insects, caught in flight or on the ground. Some small birds and reptiles.

Linnet

Status R and S; WV. Decline since the 1970s.

Habitat and nest site Breeds in young coppice (especially sweet chestnut) and young plantations, both conifer and broadleaved. Open low scrub and hedges are other breeding habitats. Both in summer and winter it feeds mainly on farmland, also on saltmarsh in winter. Roosts communally, mainly in scrub. Typically nests in a bush at heights up to 4 m.

Food and feeding site Seeds, especially those of farmland weeds. Eats fewer tree seeds than other finches. Feeds mainly on the ground.

Yellowhammer

Status R; WV.

Habitat and nest site Open scrub, young plantations (both broadleaved and conifers), young coppice (especially sweet chestnut) and hedgerows. Birds nesting in scrub and woods may feed outside these habitats probably mainly on farmland. Roosts in thick hedges or scrub. The nest is well hidden on the ground or low in the field layer.

Food and feeding site Seeds, especially of grasses, with many insects in summer. Young are fed mainly on insects. Gleans insects from bushes and trees; takes seeds on the ground.

Cirl Bunting

Status R (M). Major decline since the 1960s. Absent from Ireland.

Habitat and nest site Rare breeder now confined to south-west Britain where it breeds on farmland and in scrub. Formerly widespread in scrub habitats in southern and central England, especially on downland. In winter, it feeds mainly on farmland. Its nest site is similar to that of Yellowhammer but tends to be slightly higher, although still within 1 m of the ground.

Food and feeding site Small weed seeds. Young are fed mainly on insects. Feeding sites are similar to those of Yellowhammer.

Reed Bunting

Status R; WV. Declined mid-1970s–80s.

Habitat and nest site Main habitats are bogs and marshes with a prolific field layer, though the species expanded into farmland, open scrub and young plantations (including conifers) in 1950–60s. Some withdrawal to its wetland habitats since then. In Ireland, it has long used dry scrubby habitats. The nest is well hidden on the ground or in low vegetation within 1 m of the ground.

Food and feeding site Small grass and herb seeds, with many insects, especially in summer. Young are fed on insects. It takes more insects in winter than the above buntings but its feeding sites are similar.

PART C. Species that will nest in woodland but which feed mainly in surrounding open habitats. Species that make much use both of woodland and open habitats for feeding (e.g. Sparrowhawk, Wryneck, Green Woodpecker) have been listed in PART A.

Grey Heron

Status R; WV.

Habitat and nest site Uses woodland only for nesting. English colonies are typically in mature woods, or clumps of trees; many colonies are in oak and conifers, with nests placed in the tree canopy up to 25 m. In northern Britain, a higher percentage of birds nests on the ground, on cliffs or in scrub, and heronries containing a single nest are commoner. Colonies are often close to wetlands; coastal habitats are especially important in Scotland (Marquiss, 1989).

Food and feeding site Fish, amphibians, moles and worms. Feeds outside woodland, typically at the edges of all types of shallow water.

Red Kite

Status R (M); WV. Slow increase in the Welsh population. Re-introductions of birds from Spain and Scandinavia into England and Scotland commenced in 1989. Absent from Ireland.

Habitat and nest site Formerly bred in a wide range of woodland, now mainly in sessile oak in central Wales. The nest is usually placed in a main fork in the canopy, mainly at heights of 12–15 m.

Food and feeding site Live and dead mammals including sheep carrion, also birds and invertebrates. Hunts mainly over open country, including *ffridd*, sheepwalk and lowland farmland (Walters Davies & Davis, 1973).

Buzzard

Status R. Within Britain, largely restricted to the north and west, but the range is now expanding eastwards. Much rarer in Ireland than Britain.

Habitat and nest site Mature woodland, both coniferous and mixed, but generally uncommon in extensive upland conifer plantations. Nests in the canopy (5–20 m), though in upland habitats often nests on crags or even on the ground.

Food and feeding site Rabbits, small mammals, birds and earthworms. Feeds more in open habitats than in woodland.

Golden Eagle

Status R. Largely confined to Scotland. Absent from Ireland.

Habitat and nest site Though in Scotland almost exclusively a bird of open uplands, in Scandinavia the bird frequently lives in extensive conifer forest. Tjernberg (1983) records that 75% of 246 Swedish nest localities were in trees, typically ancient pines of average age 335 years, in stands of more than 150 years old. A few pairs in Scotland do nest in open pine forest, occasionally in birch. Nests in the canopy.

Food and feeding site Large vertebrates, dead or alive. Hunts mainly over open country, occasionally within open wooded areas.

Osprey

Status T (M). Increasing in Scotland since recolonisation in 1954. Absent from Ireland.

Habitat and nest site Coniferous forest close to lakes or large rivers, which is the typical habitat in Scotland and Scandinavia, or in more open country with scattered trees. Nests in the canopy of an often isolated large tree, usually a conifer. Elsewhere in its range the Osprey will nest on small rocky islands, or specially provided platforms, even colonially around estuaries.

Food and feeding site Fish, taken by plunge-diving.

Kestrel

Status R; WV.

Habitat and nest site A bird of open-country which will nest in various woodland types, and usually close to open feeding habitat. Nests in tree-holes, old nests and dreys, or cliff ledges. Uses nest boxes.

Food and feeding site Small mammals and birds. Hunts over open country, including young plantations.

Merlin

Status R; WV. Long-term decline. Absent from south-east Britain.

Habitat and nest site Upland bird, mostly nesting on the ground on open moors. A few nest in broadleaved trees e.g. rowans. Merlins will nest in young conifer plantations (Watson, 1979), but recently the bird has also started nesting in trees at the edges of mature spruce plantations (Parr, 1991; Little & Davison, 1992). Tree-nesters use old birds' nests in a fork or canopy (5–15 m).

Food and feeding site   Mainly small birds which are hunted over open country. Merlins do hunt over young afforestation and restocks but the importance of these areas for feeding is unknown (S.J. Petty, personal communication).

Hobby
Status   T. Increasing. Largely confined to central and southern England. Absent from Ireland.
Habitat and nest site   Will nest in isolated trees, and uses virtually all types of woodland from copses to large plantations. Many pairs nest on farmland and heathland. Nests in a fork or canopy in an old birds' nest (8–20 m).
Food and feeding site   Small birds, bats and large insects are hunted over open country.

Stock Dove
Status   R; WV. Recovering from decline in the 1950s. Absent from north-west Scotland.
Habitat and nest site   Includes woodland, parks and well-timbered farmland. It can also use coastal habitats. In eastern Europe, it is associated more strongly with coniferous forest than with deciduous. When nesting in woodland it nests in tree-holes up to some 15 m above ground. It also nests in old buildings, cliffs and rabbit burrows.
Food and feeding site   Herbivorous: seeds, buds and leaves. Feeds on the ground, usually outside woodland. In central Europe, however, the species will feed on the ground within woodland.

Turtle Dove
Status   T. Decline since the 1970s. Largely confined to south-east Britain. Has bred sporadically in Ireland.
Habitat and nest site   Thick scrub, thicket-stage plantations (including conifers), mid-coppice growth and woodland edge. Nests in a bush or low tree, 1–5 m, mostly less than 2 m above the ground.
Food and feeding site   Mainly seeds and fruits which are mostly taken on the ground. Feeds mainly outside scrub and woodland.

Long-eared Owl
Status   R; WV.
Habitat and nest site   Three main habitats: (1) mature coniferous woodland (e.g. pine plantations in East Anglia); (2) shelterbelts and small mature woods (especially in northern Britain); (3) broadleaved fen woodland, including sallow and thorn scrub (especially in southern Britain). Nests in an old nest or drey in the canopy up to 20 m. Will take to artificial nests. Winter roosts are often in thick scrub (for a comparison of Long-eared and Short-eared Owl roost habitats see Kemp (1982)).
Food and feeding site   Small mammals and birds. Probably hunts mainly over open country (including rides and young plantations). Takes most food from the ground.

Mistle Thrush
Status   R; WV.
Habitat and nest site   Wide range of woodland and lightly wooded country, though in central and eastern Europe it is a bird of coniferous forest and in southern Europe of sub-alpine conifers. Avoids extensive forest but requires

some trees. The nest is typically placed in a fork or on the bough of a tree (3–10 m).
Food and feeding site Invertebrates, which are hunted on the ground. Wide range of berries; defends berry stocks on individual trees, especially on hollies (Snow & Snow, 1984). Feeds largely outside woodland.

Magpie
Status R. Major increase but remains relatively scarce in much of Scotland.
Habitat and nest site Now occurs in virtually all habitats offering suitable nest sites in the crowns of trees or bushes. Often chooses a nest site in a dense thorn bush. Birds nesting inside woodland appear to feed mainly on surrounding open land.
Food and feeding site Omnivorous (Birkhead, 1991). Invertebrates are the main food of nestlings and adults in summer. Much vegetable matter is taken in winter. Will eat small vertebrates, carrion and birds' eggs. Feeds mainly on the ground in short grass. Hoards food.

Jackdaw
Status R; WV.
Habitat and nest site Woodland with mature trees, parkland, urban areas and cliffs. Woodland-nesters feed mainly on surrounding land. Nests gregariously, usually in holes in trees or rocks. Tree nests up to 15 m.
Food and feeding site Omnivorous. Mainly invertebrates, cereals and weed seeds. Feeds mainly on the ground, outside woodland, often in flocks. Will feed on caterpillars within woodland when these are abundant.

Rook
Status R; WV. Declined from the 1950s to 1970s, now partially recovered.
Habitat and nest site Tenuously a woodland bird. Some nesting colonies are in woods though birds feed almost entirely outside woodland. Nests in the canopy, most frequently of hedgerow trees or small copses.
Food and feeding site Earthworms and cereal grain. Wide range of other invertebrates. Feeds mainly on ground though will take caterpillars from trees when these are abundant.

Carrion/Hooded Crow
Status R; WV. Major increase.
Habitat and nest site Very wide range of habitats. Birds nesting in woodland feed mainly outside. Nests in the canopy.
Food and feeding site Omnivorous. Invertebrates, small vertebrates, carrion, seeds, grain etc. Feeds mainly on the ground, often in taller grass than Jackdaw or Rook. Will also feed in trees.

Starling
Status R; WV. Decline since the 1970s.
Habitat and nest site Wide range of habitats including woodland. Feeds mainly on farmland. Highly gregarious at all times of year. Nests in holes up to about 15 m. Uses nest boxes readily.
Food and feeding site Wide range of food. Soil invertebrates (e.g. leatherjackets) are especially important. Also takes seeds and fruits. Feeds mainly on the ground.

Tree Sparrow

Status R. Decline since the 1970s may be the latest of several fluctuations since the mid-1800s (Summers-Smith, 1989). Much scarcer in Ireland than in Britain.

Habitat and nest site In Britain, a bird of arable farmland with scattered woods and trees. Will nest in large woods when the population is high, though feeds on surrounding farmland. In Ireland and elsewhere in Europe, the species occurs in villages which it hardly ever does in Britain (Summers-Smith, 1988). British birds nest mainly in tree-holes up to 8 m. Often uses nest boxes.

Food and feeding site Insects and seeds in summer; seeds in winter. Nestlings are fed on insects, many of which are caught in tree foliage. Otherwise it feeds mainly on the ground.

PART D. Species that are essentially incidental breeders in scrub or woodland, including those that are predominantly winter visitors. Ring-necked Parakeet and Serin could arguably be added to this list but I have excluded them on the grounds that in Britain they appear to depend on large trees in parkland or gardens, rather than woods. Waxwing, an irruptive winter visitor which will feed on berries in scrub, has also been excluded because it is mainly associated with gardens and suburban areas in Britain.

Mandarin

Status R (I). Slowly expanding. Largely confined to southern Britain. Absent from Ireland.

Habitat and nest site Still or slow-flowing waters bordered by woodland or dense shrubs. Nests in a tree-hole (0–10 m). Will use nest boxes.

Food and feeding site Omnivorous. Seeds and nuts, including acorns and beechmast in winter. Many insects in summer. Takes food mainly from the water surface and by up-ending.

Mallard

Status R; WV.

Habitat and nest site Essentially associated with shallow wetlands but has adapted to a wide range of habitats. In the breeding season, it is often found at some distance from water and will nest within woodland. Nests in woodland can be on the ground, usually in dense cover, in large tree cavities, or in the forks of trees, especially pollards.

Food and feeding site Omnivorous. Takes a wide range of vegetable foods and many insects in summer, mainly from water surface or by up-ending.

Goldeneye

Status R (M); WV. Increasing since colonisation of Inverness-shire in the 1970s. Absent from Ireland.

Habitat and nest site Northern lakes and rivers bordered by coniferous forest or birch. Nests in a tree-hole or nest box (1–5 m).

Food and feeding site Aquatic invertebrates, fish fry and some vegetable matter. Feeds by diving to depths of 4 m.

Goosander

Status   R; WV. Continued increase. Virtually absent from Ireland.
Habitat and nest site   Uses riparian woodland or trees for nesting. Nests in a tree-hole up to 8 m.
Food and feeding site   Mostly fish.

Montagu's Harrier

Status   T (M). Has bred sporadically in Ireland.
Habitat and nest site   Rare lowland raptor that occurs mostly in arable farmland but has nested in young conifer plantations. Nests on the ground.
Food and feeding site   Similar to Hen Harrier.

Red-legged Partridge

Status   R (I). Virtually absent from Ireland. Distribution is greatly influenced by hand-rearing.
Habitat and nest site   Primarily a bird of farmland but uses the young stages of some lowland, sandy conifer forests. For example, in Thetford Forest it is a common bird in young pine plantations where it actually nests among the trees. Nests on the ground.
Food and feeding site   Mainly herbivorous, including many seeds. Insects are an important part of the diet of chicks. Though the bird may perch high, it feeds entirely on the ground.

Golden Plover

Status   R; WV. The British wintering population is substantially larger than the breeding population.
Habitat and nest site   A breeding bird of open moorland but it occasionally breeds in the youngest stages of first-generation, upland conifer plantations. Nests on the ground.
Food and feeding site   Invertebrates taken from the ground.

Snipe

Status   R; WV.
Habitat and nest site   In Britain, this species does not qualify in any way as a woodland bird but in Poland I have found it displaying in open damp woodland, especially of birch. Nests on the ground.
Food and feeding site   Invertebrates which are taken by probing in the ground.

Curlew

Status   R; WV.
Habitat and nest site   Primarily a breeding bird of moorland, heathland and rough grassland but it sometimes breeds in young conifer plantations, mainly in the uplands. At present it is very rare on restocks. Nests on the ground.
Food and feeding site   Invertebrates taken from the ground.

Collared Dove

Status   R. The colonisation of Britain, which started in the 1950s, is now complete.
Habitat and nest site   Occasionally nests at the edge of woodland but is mainly associated with man. The nest is usually situated in a tree, often a conifer, up to 20 m.
Food and feeding site   Mainly grain and seeds, which are taken on the ground.

Barn Owl
Status R. Marked decline since the 1950s.
Habitat and nest site Not really a woodland bird, though it will nest at the
woodland edge and within young plantations where suitable nest sites exist.
Nests in tree-holes or buildings. Will use special nest boxes.
Food and feeding site Mainly rodents. Feeds over open country, sometimes
young plantations, but mainly farmland.

Little Owl
Status R (I). Introduced in the 1800s. Absent from most of Scotland and
from Ireland.
Habitat and nest site Most typical of farmland, though it does breed in open
woods and at the woodland edge. Mainly nests in tree-holes up to 8 m.
Food and feeding site Large insects and other invertebrates, with some small
birds and mammals. A 'sit and wait' hunter, catching much prey on the
ground.

Skylark
Status R; WV. Recent decline on farmland.
Habitat and nest site Essentially a bird of farmland and grassland habitats.
However, it commonly breeds in very young plantations with a grassy field
layer. Nests on the ground.
Food and feeding site Seeds and insects which are taken on the ground.

Meadow Pipit
Status R; WV.
Habitat and nest site The majority of British breeding Meadow Pipits use
moorland, heathland and other open semi-natural habitat, but the bird also
breeds in some young conifer plantations. There are notable regional differences. For example, it is absent or a rare breeder in plantations in Thetford
Forest and the Forest of Dean, but present in many northern forests. The
species is perhaps more typical of new afforestation than of second-generation
plantations. A common wintering bird in open areas within some forests, for
example in the restocks of Thetford Forest Meadow Pipit is probably the most
abundant bird. Nests on the ground.
Food and feeding site Insects and seeds taken on the ground.

Ring Ouzel
Status S. Decreasing. Far scarcer in the Irish than in the British uplands.
Habitat and nest site During the breeding season it is an upland bird, sometimes using moorland with scattered trees and bushes, but commonly breeding
above the treeline and in rocky areas lacking any trees. In Europe (e.g. Alps,
Carpathians), it breeds within mountain conifer forest just below the tree
line, and in sub-montane scrub just above the tree line. Possibly the latter
was its primeval habitat in Britain. On passage it will occur in scrub or
wherever berries are available. Regularly uses downland scrub in southern
England. Usually nests close to the ground, often in heather, but occasionally
in a tree or bush. Birds living in European mountain forests typically nest in
conifer trees in a manner similar to Blackbird.
Food and feeding site Insects and worms in the breeding season; berries at
other times. Feeds mainly on the ground, except when taking berries from
bushes and shrubs.

Fieldfare
Status   R or S (M); mainly WV. A rare breeder. Absent from Ireland. Widespread and abundant in winter.
Habitat and nest site   Will use a wide range of scrub, woodland edges, parkland, gardens and orchards. Often nests colonially (though apparently not in Britain). Often the nest is placed in an exposed site in a bush or tree. Roosts in flocks, often in scrub.
Food and feeding site   Similar to Blackbird, though feeds gregariously in open fields and on berries in hedges and scrub.

Cetti's Warbler
Status   R (M). Colonised Britain in the 1970s. Local populations in south and east England. Absent from Ireland.
Habitat and nest site   Wetlands usually with dense scrub or damp woodland. The nest can be in low swamp vegetation or in thick bushes up to about 1 m above ground.
Food and feeding site   Insects. Feeds on, or close to, the ground in low swampy herbage and bushes.

Sedge Warbler
Status   T. Major declines in the 1970s–80s.
Habitat and nest site   Though essentially a bird of marsh and swamp habitats, it makes considerable use of scrub, for breeding and feeding, sometimes away from water. Nests in the field layer (<1 m) or in a bush (up to 4 m).
Food and feeding site   Insects. Feeds in wet swamp and drier fen vegetation including scrub.

Marsh Warbler
Status   T (M). Declined since the 1950s almost to extinction. Absent from Ireland.
Habitat and nest site   A wetland bird that breeds both in rank marsh and bushy habitats, formerly including osier beds. Nests in low dense foliage (<1 m),
Food and feeding site   Insects. Feeds both in marshy herbage and bushes.

Reed Warbler
Status   T. Virtually absent from Ireland.
Habitat and nest site   Closely associated with beds of *Phragmites* reed, though often breeds close to bushes and scrub. The nest is typically suspended in reeds (<2 m) but can be located within scrub (up to 5 m).
Food and feeding site   Insects. Feeds heavily in bushes and scrub, especially willow.

Great Grey Shrike
Status   WV. Believed to be decreasing.
Habitat and nest site   A scarce winter visitor occurring in young conifer plantations, open scrub and heathland.
Food and feeding site   Large invertebrates, small mammals and small birds. Hunts from an exposed perch.

Raven

Status R. Overall, the British population has probably declined in the last 20 years. Absent from south-east Britain.

Habitat and nest site Tenuously a woodland bird. In Britain, it rarely nests within woodland though elsewhere in Europe it does so more frequently. Often, however, it nests in the canopy of isolated trees or clumps of trees.

Food and feeding site Omnivorous: carrion, small vertebrates, a wide range of invertebrates, seeds etc. It feeds mainly on the ground outside woodland.

Brambling

Status WV. Spasmodic breeding records from northern and upland Britain.

Habitat and nest site The ecological equivalent of Chaffinch in boreal birch and far northern conifer forests. In winter, it feeds on farmland in mixed flocks of finches and buntings, occurring in woods in years of good beechmast crops. Roosts communally in young conifer plantations, bramble and scrub.

Food and feeding site Insects in summer, seeds in winter. More dependent on beechmast than Chaffinch and its distribution shifts from one winter to the next accordingly. Seeds are taken on the ground.

Goldfinch

Status R and S.

Habitat and nest site Tenuously a bird of woodland: breeds mainly in gardens, villages and parks, sometimes in hedgerows. Will feed in open woodland, such as young coppice with a profuse field layer. Roosts in a variety of sites but especially holly, ivy and other evergreens in winter. The nest site is in a bush or in a fork within the tree canopy, up to 15 m.

Food and feeding site Seeds, especially of Compositae (thistles, burdocks, etc.). Some tree seeds in winter especially pine, birch, alder. Sitka spruce seeds are also eaten (Shaw & Livingstone, 1991). Takes seed from the heads of standing thistles, teasels etc. also from the ground and the tree canopy.

Scarlet Rosefinch

Status T (M). Winters in southern Asia. Recent rare colonist. Absent from Ireland.

Habitat and nest site In Europe, it breeds mainly in lowland scrub or thickets of developing woodland, though not all British sites fit this description. Nests low in a bush, typically within 2 m of the ground.

Food and feeding site Eats seeds, buds and young leaves. Feeds on the plants rather than on the ground.

Corn Bunting

Status R; WV. Major decrease since the early 1970s. Now virtually absent from Ireland.

Habitat and nest site Essentially a bird of arable farmland. Open scrub on downland is an occasional, but atypical, breeding habitat. In winter, however, it roosts gregariously in reedbeds or thick scrub. Nests on the ground.

Food and feeding site Seeds of cultivation which are taken on the ground. Rarely, if ever, feeds in scrub.

APPENDIX 2

Woodland and scrub bird species breeding in mainland Europe but not in Britain: a summary of their distribution, breeding habitats and nest sites

Species that are essentially boreal or Mediterranean in their distribution are not included, though these are listed in footnotes. Species, such as Common Crane, which occasionally breed in woodland are not included. Codes for migrant status: **T**, tropical migrants wintering mainly south of Sahara (* indicates a species that overwinters in tropical Asia); **S**, short-distance migrants wintering either in western Europe or in the Mediterranean region; **R**, resident. Codes for main food during breeding season: **V**, vertebrate prey; **I**, invertebrates; **H**, herbivorous. Major sources include, Voous (1960), Harrison (1982), Tomiałojć & Wesołowski (1990).

Short-toed Eagle (T; V)
European distribution   Eastern/central Europe and Mediterranean, with outliers in central France.
Habitat and nest site   Open woods in the south, more dense woodland in the north, but feeds in open habitat. Typically nests in trees, often at the forest edge.

Lesser Spotted Eagle (T; V)
European distribution   Central Europe.
Habitat and nest site   Mature mixed forest and broadleaved forest. Feeds in open habitats. Nests in the canopy, often near the forest edge.

Spotted Eagle (T and S; V)
European distribution Eastern to central Europe.
Habitat and nest site Similar to Lesser Spotted Eagle, but will nest in more open and swampy woodland. Nests in the canopy.

Booted Eagle (T; V)
European distribution Similar to Short-toed Eagle.
Habitat and nest site Open or fragmented woodland and the forest edge. Mainly nests in trees, but also on cliffs.

Hazel Grouse (R; H)
European distribution Northern and central Europe, including the Alps as far west as the Belgian Ardennes.
Habitat and nest site Closed coniferous or mixed forest. Characteristic of both boreal and mountain forests. Uses treefall areas in central Europe. Nests on the ground.

Green Sandpiper (T and S; I)
European distribution Boreal Europe, south to Poland.
Habitat and nest site Swampy forest and forest bogs. Lays eggs in old tree nests of, for example, thrushes.

Scops Owl (T; I)
European distribution Southern and far eastern Europe.
Habitat and nest site Open woodland, parks and villages. Nests in tree-holes.

Eagle Owl (R; V)
European distribution Complex. Widespread from boreal to the Mediterranean with scattered populations in central Europe.
Habitat and nest site Open and closed woodland, rocky areas. Nests in tree-holes, old nests, cliff ledges and on the ground.

Pygmy Owl (R; V and I)
European distribution Boreal Europe, south to Poland with outposts further south, including the Alps.
Habitat and nest site Spruce forest. Nests in tree-holes.

Tengmalm's Owl (R; V)
European distribution Similar to Pygmy Owl.
Habitat and nest site Spruce forest. Nests in tree-holes.

Roller (T; I)
European distribution Eastern and southern Europe.
Habitat and nest site Open woodland or scattered trees. Nests in tree-holes.

Grey-headed Woodpecker (R; I)
European distribution Widespread from the far east of Europe to western France, but not Mediterranean or boreal.
Habitat and nest site Broadleaved and mixed woodland, often fairly open

stands. Frequents forest habitats more frequently than Green Woodpecker but is similar to that species in feeding on the ground. Excavates its nest hole usually in a broadleaved tree.

Black Woodpecker (R; I)
European distribution Widespread except for parts of extreme western Europe and the Mediterranean.
Habitat and nest site Mature forest; often associated with conifers though it does not necessarily prefer these to broadleaves. Excavates a large nest hole in a tree trunk which subsequently offers a nest/roost site for several other bird species and mammals.

Syrian Woodpecker (R; I)
European distribution South-east Europe, but spreading northwards.
Habitat and nest site Open woodland, gardens and parks. Less strongly associated with woodland than other European woodpeckers. Excavates its nest hole in a tree.

Middle Spotted Woodpecker (R; I)
European distribution Broadly similar to Grey-headed Woodpecker.
Habitat and nest site Mature broadleaved forest with much dead wood. Excavates its nest hole typically in the trunk of a live, broadleaved tree.

White-backed Woodpecker (R; I)
European distribution Mainly eastern Europe, Balkans and southern Scandinavia.
Habitat and nest site Mature broadleaved forest with an abundance of dead and decaying trees. Excavates its nest hole high in the tree trunk.

Three-toed Woodpecker (R; I)
European distribution Boreal Europe, south to Poland with outposts in the Alps and the Carpathians.
Habitat and nest site Mature spruce and mixed forest with much dead wood. Characteristic of both boreal and mountain forests. Excavates its nest hole, typically in the trunk of a conifer.

Thrush Nightingale (T; I)
European distribution Central and eastern Europe.
Habitat and nest site Forest edge and scrub, often swampy. Nests are placed near the ground in dense vegetation.

Bluethroat (S; I)
European distribution Widespread, but absent from much of southern Europe.
Habitat and nest site Very swampy scrub and woodland. Nests on or near the ground in dense vegetation.

River Warbler (T; I)
European distribution Eastern Europe.
Habitat and nest site Swampy forest edge. Nests on the ground in dense field layer.

Icterine Warbler (T; I)
European distribution Widespread, except for the Mediterranean region.
Habitat and nest site Woodland edge, parks and gardens. Most nests are placed in shrubs more than 1.5 m above the ground or in the canopy.

Melodious Warbler (T; I)
European distribution South-west Europe, including Iberia and Italy.
Habitat and nest site Open woodland, scattered trees or patches of bushes. Its nest site is similar to that of Icterine Warbler.

Barred Warbler (T; I)
European distribution Central and eastern Europe.
Habitat and nest site Woodland edge and scrub. Nests low in dense vegetation.

Bonelli's Warbler (T; I)
European distribution Central and south-west Europe.
Habitat and nest site Typically mature broadleaved woodland on mountain or hill slopes. Also occurs in some lowland broadleaved forests and dry pine forest.

Red-breasted Flycatcher (T*; I)
European distribution Central and eastern Europe, southern Scandinavia.
Habitat and nest site Mainly mature broadleaved and mixed woodland. Nests in tree-holes.

Collared Flycatcher (T; I)
European distribution Central and eastern Europe.
Habitat and nest site Mature broadleaved woodland. Nests in tree-holes.

Short-toed Treecreeper (R; I)
European distribution Western, central and southern Europe. Breeds in the Channel Isles.
Habitat and nest site Mainly broadleaved woodland, also parks and gardens. In central Europe, it occurs more frequently in small woods and parks than does Treecreeper. Nests behind loose bark on the tree trunk.

Nutcracker (R; H)
European distribution Southern Scandinavia, north-east Europe and mountain regions of central Europe.
Habitat and nest site Conifer forest. Nests in the canopy.

Serin (R and S; I and H)
European distribution Western, central and southern Europe.
Habitat and nest site Tenuously a woodland bird. Will use open woodland but most typical of gardens, parks and built-up areas with many trees. Occasionally breeds in Britain, but not in woodland. Nests in the canopy.

Citril Finch (R; I and H)
European distribution Central Europe and Iberia.
Habitat and nest site Open conifer forests in mountains. Nests in small trees or bushes.

Ortolan Bunting (T; I and H)
European distribution   Throughout Europe.
Habitat and nest site   Tenuously qualifies as a scrub species. Occurs in scattered trees and scrub. Nests on the ground.

Notes

1 Woodland and scrub species which have small 'marginal' populations in Britain and which are more characteristic of European habitats (see Appendix 1 for more details): Honey Buzzard, Red Kite, Goshawk, Wryneck, Fieldfare, Redwing, Cetti's Warbler, Golden Oriole, Parrot Crossbill, Scarlet Rosefinch, Cirl Bunting. Though a common breeding bird in Britain, the Swift rarely nests in tree-holes here (I am aware of this behaviour only on Speyside), though it does so more frequently in central and eastern Europe.

2 Species which are confined as breeding birds to northern conifer or birch forests (defined here as the boreal forest or taiga of Finland, Norway and Sweden; note that many of these species are also characteristic of the Siberian taiga): Hawk Owl, Ural Owl, Great Grey Owl, Waxwing, Red-flanked Bluetail, Greenish Warbler, Arctic Warbler, Siberian Tit, Siberian Jay, Brambling, White-winged Crossbill, Pine Grosbeak, Rustic Bunting.

3 Species which are confined to Mediterranean scrub and woodland; † indicates those associated with scrub habitats (garigue and maquis): Imperial Eagle, Great Spotted Cuckoo, Red-necked Nightjar, Olivaceous Warbler, Olive Tree Warbler, Marmora's Warbler †, Spectacled Warbler †, Subalpine Warbler †, Sardinian Warbler †, Cyprus Warbler †, Rüppell's Warbler †, Orphean Warbler, Sombre Tit, Cretzschmar's Bunting †.

Glossary

General sources include Peterken (1993), Polunin & Walters (1985), Mitchell & Kirby (1989), Rackham (1990)

ACIDIFICATION A process whereby the soil or water bodies are rendered increasingly acidic by precipitation of sulphur or nitrogen in weakly acidic form.

ANCIENT WOODLAND In England and Wales, denotes woodland known to have existed continuously for several hundreds of years (the usual threshold date is taken as AD 1600). Woods originating since that date are RECENT WOODLAND. Ancient woodland, therefore, includes all PRIMARY WOOD-LAND but also ancient SECONDARY WOODLAND. In Scotland, 'probable ancient sites' have been defined as ones that arose naturally before the mid-nineteenth century (Roberts et al., 1992).

BOREAL The northern zone where the climate is strongly influenced by polar air masses. Typified by extremely cold winters, warm summers, fairly low rainfall and a restricted growing season. TAIGA and tundra are the characteristic vegetation.

BRASHING Management operation in THICKET-STAGE plantations in which dead and poorly growing lower branches are removed to a height of some 2 m.

BROWSE LINE The level above ground, beyond which large herbivores are unable to browse woodland foliage.

BRYOPHYTES Mosses and liverworts.

CANOPY COVER The extent to which the woodland canopy is continuous. High canopy cover (a closed canopy) permits relatively little light to reach the ground which can result in poor development of the FIELD and SHRUB LAYERS. I use the term canopy cover to mean the amount of space between the crowns of individual trees but another aspect of canopy cover is the openness of the crown foliage itself (for example ash trees have open canopies compared with beech trees).

CANOPY LAYER That part of the forest foliage contributed by the largest trees, and which generally lies above the SHRUB LAYER. Occasionally it is possible to discern an upper and lower canopy layer. See also FOLIAGE PROFILE.

CARR A swampy woodland where typically the tree species are alder, ash and willows.

CLEANING A management operation in young plantations to cut out unwanted invasive species such as birch and honeysuckle.

CLEAR-CUTTING See CLEAR-FELLING.

CLEAR-FELL An area of recently harvested woodland. Can also be termed a *clear-cut*.

CLEAR-FELLING System of managing HIGH FOREST in which blocks of trees, typically several hectares in extent, are felled and replanted. Also known as *clear-cutting*.

COMMUNITY An assemblage of POPULATIONS of different species. Biological communities vary in the types and numbers of species they contain.

CONTINUOUS COVER Systems of managing HIGH FOREST which seek to maintain a more or less continuous canopy cover through the removal of individual trees or very small patches of trees; these include SELECTION systems.

COPPICE An ancient system of growing broadleaved trees whereby the trees regenerate from the cut stumps which are called STOOLS. The multi-stemmed trees are generally cut on short rotations, typically between 7 and 35 years depending on tree species and markets, to produce crops of poles. *Coppice with standards* has larger, single-stem trees scattered among the coppiced trees; *simple coppice* has no such trees. See also UNDERWOOD.

COUPE A felled area of woodland. An alternative word is *compartment*. See also PANEL.

DENSITY Animal abundance expressed as numbers per unit area. See also POPULATION.

DIVERSITY Sometimes diversity is used simply to mean the number of species in a defined area (SPECIES RICHNESS) but this is just one component of diversity. The other is the relative abundance of the species, i.e. their commonness or rarity. A community containing 20 species, each of similar abundance, is more diverse than one containing 20 species where 50% of the individuals belong to the single most common species and half the species are very rare.

ECOLOGICAL NICHE This concept is an expression of the requirements of a plant or animal for it to survive and reproduce. Potentially, there are many dimensions to a niche but they are of two main types: environmental conditions (e.g. climate) and resources (e.g. food, habitat structure). To a large extent, each species has its own characteristic niche.

ECOTONE A transition zone between two adjacent vegetation types. This transition vegetation may have its own distinctive characteristics and often involves some type of scrub. An example would be woodland grading through scrub into grassland or heathland. Ecotones are different to *edges* in that they represent a gradient rather than an abrupt change. See also SUCCESSION.

EDGE EFFECT This occurs where the abundance of a species increases or decreases at the interface between two different types of vegetation or habitat.

EPIPHYTE A plant which grows upon another plant e.g. ivy, honeysuckle.

ESTABLISHMENT STAGE The early stages of forest or woodland growth when the canopy is open; typically the first 10 years after planting.

FFRIDD Enclosed rough land at the lower edge of Welsh moorlands, typified by bracken and thorn scrub.

FIELD LAYER The stratum of forest vegetation growing closest to the ground (usually no more than 1 m above ground) which typically consists of grasses, ferns, herbs and bramble. See also SHRUB LAYER, CANOPY LAYER, FOLIAGE PROFILE.

FLORISTICS This refers to the plant species composition of vegetation, rather than to its physical attributes (the latter is the *structure*). For example, two floristic comparisons that can be made of different woods are: (1) the numbers of species of trees and shrubs that they contain; (2) the relative numbers of different species of trees and shrubs.

FOLIAGE HEIGHT DIVERSITY An index describing the complexity of the FOLIAGE PROFILE. High values of the index imply many layers of foliage. Abbreviated to FHD.

FOLIAGE PROFILE The vertical structure of woodland vegetation. The foliage profile of a stand can be divided into FIELD LAYER, SHRUB LAYER, and CANOPY LAYER. This profile can be depicted diagrammatically to indicate the relative development of the foliage in these various layers. See also WOODLAND STRUCTURE.

FOREST Historically a Forest, always with a capital F, was a tract of land subject to special laws for game preservation (as in *Royal Forests* and *chases*). Some of these Forests contained little woodland. The modern usage of the word is far more general, relating to any extensive tract of tree-covered land. Large areas of more or less natural tree cover are always termed forest. In Britain, the word has recently been applied specifically to large conifer plantations, such as those in the uplands.

GARIGUE A Mediterranean scrub vegetation consisting of low, scattered shrubs. This type of scrub is lower than MAQUIS.

GLADE A permanent open area within woodland, often grassy and maintained by grazing.

GROUP-FELLING System of managing HIGH FOREST in which small groups of trees, typically no larger than 0.5 ha, are felled and replanted. Sometimes the new crop is allowed to regenerate naturally.

HIGH FOREST Commercial coniferous or broadleaved forest in which trees are grown in a single-stemmed form, unlike coppice, and where they are usually grown on relatively long rotations to produce crops of timber. Methods of high forest management include CLEAR-FELLING, GROUP-FELLING, SELECTION, TWO-STOREY SYSTEMS, SHELTERWOOD.

KRUMMHOLZ Woodland at the altitudinal or exposure limits of tree growth and consisting of gnarled, stunted trees.

MAQUIS A Mediterranean scrub vegetation, taller and denser than GARIGUE, often containing many evergreen shrubs up to 3 m tall.

MIXED FOREST Ambiguous term which can be used to mean a stand composed of a mixture of any species but which is better restricted to stands containing both coniferous and deciduous trees in substantial quantity.

NATURAL REGENERATION The establishment of a new stand by allowing trees to grow from seed or suckers, rather than by planting.

NATURAL WOODLAND Stands where there is a policy of no management so that changes in the structure of the woodland arise from natural processes such as storms and the death of individual trees. In time, such stands will start to acquire the characteristics of OLD-GROWTH or PRIMEVAL FOREST.

NICHE See ECOLOGICAL NICHE.

NURSE CROP Where a mixture of trees is grown together within a stand, one fast-growing species, usually a conifer, may be grown as a nurse crop. This species will be removed at an early stage once the other trees have become well established. The purpose can be twofold – to improve growth and to provide a staggered financial return from the wood.

OLD-GROWTH This term is widely accepted in North America to denote mature PRIMARY stands largely undisturbed by man. The main attributes of these stands are: (1) an abundance of dead wood, as standing dead trees, dead limbs on living trees, and as fallen trunks and branches; (2) some individual trees grow to an extremely large size; (3) the stands will be patchy as a result of natural disturbances such as storm damage. See also NATURAL WOODLAND, PRIMEVAL FOREST.

PANEL A coppice compartment i.e. a patch of coppiced woodland where the trees are at the same age of growth. Other names include *hagg, cant, sale, fell*.

PARK An enclosed area of grazed land, usually with scattered trees, where historically the primary purpose was to retain deer. Can be regarded as an extreme form of wood-pasture.

PASSERINE A species belonging to the order Passeriformes, often termed the *perching birds*. Among families of birds classified as passerines are larks, pipits, thrushes, warblers, flycatchers, tits, crows, finches and buntings.

POINT COUNT A method of counting birds which involves remaining stationary at a particular place (a point) and counting birds for a pre-determined period, usually between 5 and 20 minutes.

POLE STAGE The later stages of growth of commercial conifer stands. Also known as the *pre-felling stage*.

POLLARD A tree that has been cut 2–3 m above ground to provide a crop of poles. Pollards are usual in WOOD-PASTURE and other places where there is a need to prevent large herbivores browsing the foliage.

POPULATION A group of individuals of the same species living within a defined area. Populations vary in their DENSITY which is determined by the balance of births, deaths, immigration and emigration. See also POPULATION SINK, POPULATION SOURCE.

POPULATION SINK An area or habitat in which the existence of a population depends on immigration because production of young is insufficient to maintain numbers.

POPULATION SOURCE An area or habitat in which a population produces a surplus of individuals which may emigrate and settle elsewhere.

PRE-FELLING See POLE STAGE.

PRE-THICKET STAGE See THICKET STAGE.

PRIMARY WOODLAND Woodland standing on land thought to have been continuously wooded and never cultivated. See also SECONDARY WOOD-LAND.

PRIMEVAL FOREST This term means much the same as OLD-GROWTH, though it is more commonly applied in Europe than in North America. *Virgin forest* has never been interfered with by man, either directly through management, or indirectly through external influences such as drainage of surrounding land or atmospheric pollution. Such situations are now so rare in temperate forests (they may be more frequent in the boreal zone) that it is preferable to use the terms OLD-GROWTH, PRIMEVAL or NATURAL.

RECENT WOODLAND Woods that are not ancient i.e. those which have become established since AD 1600 on formerly open land. See also ANCIENT WOODLAND.

REGENERATION See RESTOCKING.

RESTOCKING The process of promoting the next crop of trees after the felling of the preceding crop. Restocking may be achieved by planting, by regrowth from cut stumps, as in COPPICE, or through NATURAL REGENER-ATION.

RETENTION A stand in a commercially productive wood which has been allowed to grow on beyond the optimum commercial felling age for landscape or conservation reasons.

RIPARIAN Pertaining to rivers.

SAPROXYLIC SPECIES One depending on wood, usually dead wood, at some part of its life cycle. Most of these species are invertebrates or fungi, but birds and mammals which nest or roost in tree-holes could also be classed as saproxylic.

SECONDARY WOODLAND Woodland standing on ground which has been cleared at some time in its history and used for some other purpose than growing trees, usually agriculture. See also PRIMARY WOODLAND.

SELECTION System of managing HIGH FOREST in which trees are removed individually, with the result that the appearance of the wood is more constant than in most other high forest management systems. See also CONTINUOUS COVER.

SEMI-NATURAL WOODLAND In the context of British woodland, semi-natural stands are those composed of native trees and shrubs not thought to have been planted but which have arisen from natural regeneration or coppice regrowth (Roberts et al., 1992; Spencer & Kirby, 1992). Such stands may have been felled at various times. See also NATURAL WOODLAND.

SHELTERWOOD System of managing HIGH FOREST in which the new crop is regenerated entirely from seed trees which are left standing for several years, usually following an initial heavy thinning, until the new growth has established.

SHRUB LAYER The stratum of woody shrubs that often grows between the FIELD and the CANOPY LAYERS of forests. Typically the shrub layer occupies the zone between 1 m and 6 m above ground, though in many woods it is sparse or even completely absent. See also FOLIAGE PROFILE.

SINGLING A method of thinning old coppice which involves cutting all but one stem.

SNAG North American word for a standing dead tree.

SONGBIRD A general term for PASSERINES, usually excluding the crow family.

SPECIES RICHNESS The total number of species inhabiting a defined area or habitat.

STAND An area of woodland, often uniform in tree species, composition and structure.

STANDARD A tree grown for timber as opposed to wood.

STOCKING General term for the quantity of trees within a stand.

STOOL A cut COPPICE stump from which the coppice regrows.

SUCCESSION The sequence of changes that occurs in the vegetation of open land in the absence of heavy grazing, cultivation or cutting. *Primary succession* takes place when land is created anew, as on shingle banks in rivers. *Secondary succession* is far more common and occurs wherever factors previously suppressing the vegetation are removed or where storms or fire damage natural forests. Virtually all grassland, heathland, moorland and farmland in Britain would undergo secondary succession if grazing pressure was relaxed or farming practices abandoned. Most successions involve gradual change towards woodland, through a transitional scrub phase. The rate of change in vegetation gradually slows as the vegetation moves towards an apparently stable state which is called the *climax*.

TAIGA The northern BOREAL forests dominated by conifers and lying to the south of the tundra in Scandinavia and Russia.

TEMPERATE A climatic zone lying between the northern BOREAL zone and the Mediterranean zone. Includes Britain, Ireland, western mainland Europe and central Europe.

TERRITORY MAPPING A method of counting breeding birds which involves recording the birds, usually over several visits, on a large-scale map, in such a way that the locations and numbers of territories can be estimated.

THICKET STAGE The stage of forest growth after canopy-closure when the trees form a dense, often impenetrable, bushy growth. Especially applied to conifer forests of some 10 to 20 years of growth. The period just before, and covering, canopy-closure is often termed PRE-THICKET.

TRANSECT A method of counting birds which involves walking a route and recording birds encountered, often within some pre-determined distance from the observer.

TREE SHELTER Plastic tube in which a young tree is grown. This provides shelter both from herbivores and inclement weather. Growth rates are increased by the use of tree shelters.

TWO-STOREY SYSTEMS An approach to managing HIGH FOREST which involves planting trees beneath an existing canopy (UNDERPLANTING) which may have been thinned.

UNDERPLANTING The planting of a new crop of trees beneath an existing canopy. See also TWO-STOREY SYSTEMS.

UNDERSTOREY Alternative for SHRUB LAYER. May be used to signify the FIELD and SHRUB LAYERS combined.

UNDERWOOD Coppice growth, either growing within the wood, or cut. The regrowth of underwood is sometimes called *spring*.

WALDSTERBEN German word for the recent large-scale death of forests occurring mainly in central Europe.

WAYLEAVE Strip of land either side of powerlines in which trees are not allowed to grow. Woody regrowth is periodically cut or treated with herbicides.

WILDWOOD The name given by Oliver Rackham to the primeval forest that covered much of Britain before its destruction by man.

WINDTHROW Areas of storm-damaged woodland with a substantial number of fallen trees.

WOOD-PASTURE An ancient system of land-use in which domestic animals are grazed within woodland. The trees are often managed by pollarding.

WOODLAND STRUCTURE The physical architecture of woodland vegetation. There are two components of structure: (1) vertical structure is the distribution of the foliage between the ground and the top of the canopy; (2) horizontal structure is the spatial patchiness of the woodland in terms of the amount of permanently open space and the extent to which the woodland is fragmented into patches at different stages of growth. See also FOLIAGE PROFILE.

YOUNG-GROWTH Young stages of woodland development up to, but not beyond, canopy-closure.

APPENDIX 4

Scientific names

Birds (following Voous, 1977)

Grey Heron	*Ardea cinerea*
Mandarin	*Aix galericulata*
Mallard	*Anas platyrhynchos*
Goldeneye	*Bucephala clangula*
Goosander	*Mergus merganser*
Honey Buzzard	*Pernis apivorus*
Black Kite	*Milvus migrans*
Red Kite	*Milvus milvus*
Short-toed Eagle	*Circaetus gallicus*
Hen Harrier	*Circus cyaneus*
Montagu's Harrier	*Circus pygargus*
Goshawk	*Accipiter gentilis*
Sparrowhawk	*Accipiter nisus*
Buzzard	*Buteo buteo*
Lesser Spotted Eagle	*Aquila pomarina*
Spotted Eagle	*Aquila clanga*
Imperial Eagle	*Aquila heliaca*
Golden Eagle	*Aquila chrysaetos*
Booted Eagle	*Hieraaetus pennatus*
Osprey	*Pandion haliaetus*
Kestrel	*Falco tinnunculus*
Merlin	*Falco columbarius*
Hobby	*Falco subbuteo*
Hazel Grouse	*Bonasa bonasia*
Willow/Red Grouse	*Lagopus lagopus*
Black Grouse	*Tetrao tetrix*
Capercaillie	*Tetrao urogallus*
Red-legged Partridge	*Alectoris rufa*
Pheasant	*Phasianus colchicus*
Golden Pheasant	*Chrysolophus pictus*
Lady Amherst's Pheasant	*Chrysolophus amherstiae*
Common Crane	*Grus grus*
Stone-curlew	*Burhinus oedicnemus*
Golden Plover	*Pluvialis apricaria*
Lapwing	*Vanellus vanellus*
Snipe	*Gallinago gallinago*

Birds (Cont)

Woodcock	*Scolopax rusticola*
Curlew	*Numenius arquata*
Spotted Redshank	*Tringa erythropus*
Greenshank	*Tringa nebularia*
Green Sandpiper	*Tringa ochropus*
Wood Sandpiper	*Tringa glareola*
Stock Dove	*Columba oenas*
Woodpigeon	*Columba palumbus*
Collared Dove	*Streptopelia decaocto*
Turtle Dove	*Streptopelia turtur*
Ring-necked Parakeet	*Psittacula krameri*
Great Spotted Cuckoo	*Clamator glandarius*
Cuckoo	*Cuculus canorus*
Barn Owl	*Tyto alba*
Scops Owl	*Otus scops*
Eagle Owl	*Bubo bubo*
Hawk Owl	*Surnia ulula*
Pygmy Owl	*Glaucidium passerinum*
Little Owl	*Athene noctua*
Tawny Owl	*Strix aluco*
Northern Spotted Owl	*Strix occidentalis caurina*
Ural Owl	*Strix uralensis*
Great Grey Owl	*Strix nebulosa*
Long-eared Owl	*Asio otus*
Short-eared Owl	*Asio flammeus*
Tengmalm's Owl	*Aegolius funereus*
Nightjar	*Caprimulgus europaeus*
Red-necked Nightjar	*Caprimulgus ruficollis*
Swift	*Apus apus*
Roller	*Coracias garrulus*
Wryneck	*Jynx torquilla*
Grey-headed Woodpecker	*Picus canus*
Green Woodpecker	*Picus viridis*
Black Woodpecker	*Dryocopus martius*
Great Spotted Woodpecker	*Dendrocopos major*
Syrian Woodpecker	*Dendrocopos syriacus*
Middle Spotted Woodpecker	*Dendrocopos medius*
White-backed Woodpecker	*Dendrocopos leucotos*
Lesser Spotted Woodpecker	*Dendrocopos minor*
Three-toed Woodpecker	*Picoides tridactylus*
Woodlark	*Lullula arborea*
Skylark	*Alauda arvensis*
Tree Pipit	*Anthus trivialis*
Meadow Pipit	*Anthus pratensis*
Waxwing	*Bombycilla garrulus*
Wren	*Troglodytes troglodytes*
Dunnock	*Prunella modularis*
Robin	*Erithacus rubecula*
Thrush Nightingale	*Luscinia luscinia*
Nightingale	*Luscinia megarhynchos*

Birds (Cont)

Bluethroat	*Luscinia svecica*
Red-flanked Bluetail	*Tarsiger cyanurus*
Redstart	*Phoenicurus phoenicurus*
Whinchat	*Saxicola rubetra*
Stonechat	*Saxicola torquata*
Ring Ouzel	*Turdus torquatus*
Blackbird	*Turdus merula*
Fieldfare	*Turdus pilaris*
Song Thrush	*Turdus philomelos*
Redwing	*Turdus iliacus*
Mistle Thrush	*Turdus viscivorus*
Cetti's Warbler	*Cettia cetti*
Grasshopper Warbler	*Locustella naevia*
River Warbler	*Locustella fluviatilis*
Sedge Warbler	*Acrocephalus schoenobaenus*
Marsh Warbler	*Acrocephalus palustris*
Reed Warbler	*Acrocephalus scirpaceus*
Olivaceous Warbler	*Hippolais pallida*
Olive-tree Warbler	*Hippolais olivetorum*
Icterine Warbler	*Hippolais icterina*
Melodious Warbler	*Hippolais polyglotta*
Marmora's Warbler	*Sylvia sarda*
Dartford Warbler	*Sylvia undata*
Spectacled Warbler	*Sylvia conspicillata*
Subalpine Warbler	*Sylvia cantillans*
Sardinian Warbler	*Sylvia melanocephala*
Cyprus Warbler	*Sylvia melanothorax*
Rüppell's Warbler	*Sylvia rueppelli*
Orphean Warbler	*Sylvia hortensis*
Barred Warbler	*Sylvia nisoria*
Lesser Whitethroat	*Sylvia curruca*
Whitethroat	*Sylvia communis*
Garden Warbler	*Sylvia borin*
Blackcap	*Sylvia atricapilla*
Greenish Warbler	*Phylloscopus trochiloides*
Arctic Warbler	*Phylloscopus borealis*
Bonelli's Warbler	*Phylloscopus bonelli*
Wood Warbler	*Phylloscopus sibilatrix*
Chiffchaff	*Phylloscopus collybita*
Willow Warbler	*Phylloscopus trochilus*
Goldcrest	*Regulus regulus*
Firecrest	*Regulus ignicapillus*
Spotted Flycatcher	*Muscicapa striata*
Red-breasted Flycatcher	*Ficedula parva*
Collared Flycatcher	*Ficedula albicollis*
Pied Flycatcher	*Ficedula hypoleuca*
Long-tailed Tit	*Aegithalos caudatus*
Marsh Tit	*Parus palustris*
Sombre Tit	*Parus lugubris*
Willow Tit	*Parus montanus*

Birds (Cont)

Siberian Tit	*Parus cinctus*
Crested Tit	*Parus cristatus*
Coal Tit	*Parus ater*
Blue Tit	*Parus caeruleus*
Great Tit	*Parus major*
Nuthatch	*Sitta europaea*
Treecreeper	*Certhia familiaris*
Short-toed Treecreeper	*Certhia brachydactyla*
Golden Oriole	*Oriolus oriolus*
Red-backed Shrike	*Lanius collurio*
Great Grey Shrike	*Lanius excubitor*
Jay	*Garrulus glandarius*
Siberian Jay	*Perisoreus infaustus*
Magpie	*Pica pica*
Nutcracker	*Nucifraga caryocatactes*
Jackdaw	*Corvus monedula*
Rook	*Corvus frugilegus*
Carrion/Hooded Crow	*Corvus corone*
Raven	*Corvus corax*
Starling	*Sturnus vulgaris*
Tree Sparrow	*Passer montanus*
Chaffinch	*Fringilla coelebs*
Brambling	*Fringilla montifringilla*
Serin	*Serinus serinus*
Citril Finch	*Serinus citrinella*
Greenfinch	*Carduelis chloris*
Goldfinch	*Carduelis carduelis*
Siskin	*Carduelis spinus*
Linnet	*Carduelis cannabina*
Redpoll	*Carduelis flammea*
White-winged Crossbill	*Loxia leucoptera*
Common Crossbill	*Loxia curvirostra*
Scottish Crossbill	*Loxia scotica*
Parrot Crossbill	*Loxia pytyopsittacus*
Scarlet Rosefinch	*Carpodacus erythrinus*
Pine Grosbeak	*Pinicola enucleator*
Bullfinch	*Pyrrhula pyrrhula*
Hawfinch	*Coccothraustes coccothraustes*
Black-throated Green Warbler	*Dendroica virens*
Yellowhammer	*Emberiza citrinella*
Cirl Bunting	*Emberiza cirlus*
Ortolan Bunting	*Emberiza hortulana*
Cretzschmar's Bunting	*Emberiza caesia*
Rustic Bunting	*Emberiza rustica*
Reed Bunting	*Emberiza schoeniclus*
Corn Bunting	*Miliaria calandra*

Mammals

Fallow Deer	*Dama dama*
Fox	*Vulpes vulpes*

Mammals (Cont)

Lynx	*Lynx lynx*
Muntjac	*Muntiacus reevesi*
Rabbit	*Oryctolagus cuniculus*
Red Deer	*Cervus elaphus*
Roe Deer	*Capreolus capreolus*
Wolf	*Canis lupus*

Insects

Pine Beauty Moth	*Panolis flammea*

Plants

Alder	*Alnus glutinosa*
Alder Buckthorn	*Frangula alnus*
Ash	*Fraxinus excelsior*
Beech	*Fagus sylvatica*
Bell-heather	*Erica cinerea*
Bilberry	*Vaccinium myrtillus*
Birch	*Betula* spp.
Blackthorn	*Prunus spinosa*
Bluebell	*Hyacinthoides non-scripta*
Bracken	*Pteridium aquilinum*
Bramble	*Rubus fruticosus*
Buckthorn	*Rhamnus catharticus*
Cherry	*Prunus avium*
Cowberry	*Vaccinium vitis-idaea*
Dog's mercury	*Mercurialis perennis*
Dogwood	*Cornus sanguinea*
Douglas fir	*Pseudotsuga menziesii*
Dwarf gorse	*Ulex minor*
Elder	*Sambucus nigra*
Field maple	*Acer campestre*
Gorse	*Ulex europaeus*
Hawthorn	*Crataegus monogyna*
Hazel	*Corylus avellana*
Heather (Ling)	*Calluna vulgaris*
Holm oak	*Quercus ilex*
Honeysuckle	*Lonicera periclymenum*
Hornbeam	*Carpinus betulus*
Ivy	*Hedera helix*
Juniper	*Juniperus communis*
Larch	*Larix* spp.
Lodgepole pine	*Pinus contorta*
Norway spruce	*Picea abies*
Osier	*Salix viminalis*
Pedunculate oak	*Quercus robur*
Poplar	*Populus* spp.
Primrose	*Primula vulgaris*
Privet	*Ligustrum vulgare*
Reed	*Phragmites communis*
Rhododendron	*Rhododendron ponticum*

Plants (Cont)

Rose	*Rosa* spp.
Rowan	*Sorbus aucuparia*
Scots pine	*Pinus sylvestris*
Sea buckthorn	*Hippophaë rhamnoides*
Sessile oak	*Quercus petraea*
Silver fir	*Abies alba*
Sitka spruce	*Picea sitchensis*
Small-leaved lime	*Tilia cordata*
Sweet chestnut	*Castanea sativa*
Sycamore	*Acer pseudoplatanus*
Traveller's joy	*Clematis vitalba*
Willow	*Salix* spp.
Wayfaring tree	*Viburnum lantana*
Woodland hawthorn	*Crataegus laevigata*
Yew	*Taxus baccata*

REFERENCES

Able, K.P. & Noon, B.R. (1976). Avian community structure along elevational gradients in the Northeastern United States. *Oecologia* **26**, 275–294.

Adams, M.W. & Edington, J.M. (1973). A comparison of song-bird populations in mature coniferous and broadleaved woods. *Forestry* **46**, 191–201.

Adriaensen, F. & Dhondt, A.A. (1990*a*). Territoriality in the Continental European Robin *Erithacus rubecula rubecula*. *Ardea* **78**, 459–465.

Adriaensen, F. & Dhondt, A.A. (1990*b*). Population dynamics and partial migration of the European Robin (*Erithacus rubecula*) in different habitats. *Journal of Animal Ecology* **59**, 1077–1090.

Alatalo, R.V. & Lundberg, A. (1983). Laboratory experiments on habitat separation and foraging efficiency in Marsh and Willow Tits. *Ornis Scandinavica* **14**, 115–122.

Alerstam, T., Nilsson, S.G. & Ulfstrand, S. (1974). Niche differentiation during winter in woodland birds in southern Sweden and the island of Gotland. *Oikos* **25**, 321–330.

Alexander, I. & Cresswell, B. (1990). Foraging by Nightjars *Caprimulgus europaeus* away from their nesting areas. *Ibis* **132**, 568–574.

Anderson, P. & Radford, E. (1992). *A Review of the Effects of Recreation on Woodland Soils, Vegetation and Fauna*. English Nature Research Report 27. Peterborough: English Nature.

Andrén, H. & Angelstam, P. (1988). Elevated predation rates as an edge effect in habitat islands: experimental evidence. *Ecology* **69**, 544–547.

Andrews, J. (1990). Management of lowland heathlands for wildlife. *British Wildlife* **1**, 336–346.

Angelstam, P., Lindström, E. & Widén, P. (1984). Role of predation in short-term population fluctuations of some birds and mammals in Fennoscandia. *Oecologia* **62**, 199–208.

Anon. (1989). Goshawk breeding habitat in lowland Britain. *British Birds* **82**, 56–67.

Askins, R.A., Philbrick, M.J. & Sugeno, D.S. (1987). Relationship between the regional abundance of forest and the composition of forest bird communities. *Biological Conservation* **39**, 129–152

Avery, M.I. & Leslie, R. (1990). *Birds & Forestry*. London: Poyser.

Avery, M.I., Winder, F.W.R. & Egan, V. (1989). Predation on artificial nests adjacent to forestry plantations in northern Scotland. *Oikos* **55**, 321–323.

Baillie, S.R. & Peach, W.J. (1992). Population limitation in Palaearctic-African migrant passerines. *Ibis* **134** (suppl.1), 120–132.

220

Bain, C. & Bainbridge, I. (1988). A better future for our native pinewoods? *RSPB Conservation Review* **2**, 50–53.

Baines, D. (1990). Black Grouse densities and habitat requirements. *Game Conservancy Review of 1989*, 136–138.

Baines, D. (1991). Factors affecting Black Grouse breeding success. *Game Conservancy Review of 1990*, 159–161.

Baines, D. & Sage, R. (1992). Capercaillie decline and caterpillar abundance in Highland pine woods. *Game Conservancy Review of 1991*, 104–105.

Baker, R.R. (1993). The function of post-fledging exploration: a pilot study of three species of passerines ringed in Britain. *Ornis Scandinavica* **24**, 71–79.

Bamford, R. (1985). Factors affecting the songbird communities of young conifer plantations. *Nature in Wales* **4**, 82–87.

Batten, L.A. (1973). The colonisation of England by the Firecrest. *British Birds* **66**, 159–166.

Batten, L.A. (1976). Bird communities of some Killarney woodlands. *Proceedings of the Royal Irish Academy* **76**, 285–313.

Batten, L.A., Bibby, C.J., Clement, P., Elliott, G.D. & Porter, R.F. (eds.) (1990). *Red Data Birds in Britain*. London: Poyser.

Bayes, K. & Henderson, A. (1988). Nightingales and coppiced woodland. *RSPB Conservation Review* **2**, 47–49.

Beven, G. (1959). The feeding sites of birds in dense oakwood. *London Naturalist* **38**, 64–73.

Beven, G. (1964). The feeding sites of birds in grassland with thick scrub. *London Naturalist* **43**, 86–109.

Beven, G. (1976). Changes in breeding bird populations of an oak-wood on Bookham Common, Surrey, over twenty-seven years. *London Naturalist* **55**, 23–42.

Bibby, C.J. (1978). A heathland bird census. *Bird Study* **25**, 87–96.

Bibby, C.J. (1979a). Breeding biology of the Dartford Warbler *Sylvia undata* in England. *Ibis* **121**, 41–52.

Bibby, C.J. (1979b). Foods of the Dartford Warbler *Sylvia undata* on southern English heathland (Aves:Sylviidae). *Journal of Zoology, London.* **188**, 557–576.

Bibby, C.J. (1989). A survey of breeding Wood Warblers *Phylloscopus sibilatrix* in Britain, 1984–1985. *Bird Study* **36**, 56–72.

Bibby, C.J., Aston, N. & Bellamy, P.E. (1989b). Effects of broadleaved trees on birds of upland conifer plantations in north Wales. *Biological Conservation* **49**, 17–29.

Bibby, C.J., Bain, C.G. & Burges, D.J. (1989a). Bird communities of highland birchwoods. *Bird Study* **36**, 123–133.

Bibby, C.J. & Etheridge, B. (1993). Status of the Hen Harrier *Circus cyaneus* in Scotland in 1988–89. *Bird Study* **40**, 1–11.

Bibby, C.J., Phillips, B.N. & Seddon, A.J.E. (1985). Birds of restocked conifer plantations in Wales. *Journal of Applied Ecology* **22**, 619–633.

Bibby, C.J. & Robins, M. (1985). An exploratory analysis of species and community relationships with habitat in western oak woods. In *Bird Census and Atlas Studies* (ed. K. Taylor, R.J. Fuller & P.C. Lack), pp. 255–264. Proceedings VIII International Conference on Bird Census and Atlas Work. Tring: British Trust for Ornithology.

Bijlsma, R.G. (1986). Occurrence and breeding biology of the Honey Buzzard *Pernis apivorus* on the S.W. Veluwe and in the S.E. Achterhoek. (English title.) *Limosa* **59**, 61–66.

Bijlsma, R.G. (1988). The Goshawk *Accipiter gentilis* in The Netherlands in the 20th century. (English title.) *Limosa* **61**, 133–136.

Bijlsma, R.G. (1993). *Ecologische Atlas van de Roofvogels in Nederland*. Haarlem: Schuyt.

Bilcke, G. (1984). Seasonal changes in habitat use of resident passerines. *Ardea* **72**, 95–99.

Bilcke, G., Mertens, R., Jeurissen, M. & Dhondt, A.A. (1986). Influences of habitat structure and temperature on the foraging niches of the Pariform guild in Belgium during winter. *Le Gerfaut* **76**, 109–129.

Birkhead, T.R. (1991). *The Magpies – the Ecology and Behaviour of Black-billed and Yellow-billed Magpies*. London: Poyser.

Blake, J.G. & Hoppes, W.G. (1986). Influence of resource abundance on use of tree-fall gaps by birds in an isolated woodlot. *Auk* **103**, 328–340.

Blondel, J. (1981). Structure and dynamics of bird communities in Mediterranean habitats. In *Mediterranean-Type Shrublands* (ed. F. di Castri, D.W. Goodall & R.L. Specht), pp. 361–385. Amsterdam: Elsevier.

Boddy, M. (1991). Some aspects of frugivory by bird populations using coastal dune scrub in Lincolnshire. *Bird Study* **38**, 188–199.

Bossema, L. (1979). Jays and oaks: an eco-ethological study of a symbiosis. *Behaviour* **70**, 1–117.

Bowden, C.G.R. (1990). Selection of foraging habitats by Woodlarks (*Lullula arborea*) nesting in pine plantations. *Journal of Applied Ecology* **27**, 410–419.

Bowden, C. & Hoblyn, R. (1990). The increasing importance of restocked conifer plantations for Woodlarks in Britain: implications and consequences. *RSPB Conservation Review* **4**, 26–31.

Bradshaw, A.D. (1983). The reconstruction of ecosystems. *Journal of Applied Ecology* **20**, 1–17.

Bremer, P. (1980). Broedvogels van het Kuinderbos. *Vogeljaar* **28**, 287–291.

British Ornithologists' Union. (1992). *Checklist of Birds of Britain and Ireland*, 6th edn. Tring: British Ornithologists' Union.

Burgess, N., Evans, C. & Sorensen, J. (1990). Heathland management for Nightjars. *RSPB Conservation Review* **4**, 32–35.

Cadbury, C.J. (1989). What future for lowland heaths in southern Britain? *RSPB Conservation Review* **3**, 61–67.

Campbell, B. & Ferguson-Lees, I.J. (1972). *A Field Guide to Birds' Nests*. London: Constable.

Cannell, M.G.R. (1990). Forests. In *The Greenhouse Effect and Terrestrial Ecosystems of the UK* (ed. M.G.R. Cannell & M.D. Hooper), pp. 24–26. London: Her Majesty's Stationery Office.

Cannell, M.G.R., Grace, J. & Booth, A. (1989). Possible impacts of climatic warming on trees and forests in the United Kingdom: a review. *Forestry* **62**, 337–364.

Casey, D. & Hein, D. (1983). Effects of heavy browsing on a bird community in deciduous forest. *Journal of Wildlife Management* **47**, 829–836.

Catt, D.C., Dugan, D., Green, R.E., Moncrieff, R., Moss, R., Picozzi, N., Summers, R.W. & Tyler, G.A. (1994). Collisions against fences by woodland grouse in Scotland. *Forestry* **67**, 105–118.

Cayford, J.T. (1993). *Black Grouse and Forestry: Habitat Requirements and Management*. Forestry Commission Technical Paper 1. Edinburgh: Forestry Commission.

Cayford, J. & Hope Jones, P. (1989). Black Grouse in Wales. *RSPB Conservation Review* **3**, 79–81.

Clarke, R. & Watson, D. (1990). The Hen Harrier *Circus cyaneus* winter roost survey in Britain and Ireland. *Bird Study* **37**, 84–100.

Collins, S.L. (1983). Geographic variation in habitat structure of the Black-throated Green Warbler (*Dendroica virens*). *Auk* **100**, 382–389.

Colquhoun, M.K. & Morley, A. (1943). Vertical zonation in woodland bird communities. *Journal of Animal Ecology* **12**, 75–81.

Connor, E.F. & McCoy, E.D. (1979). The statistics and biology of the species-area relationship. *American Naturalist* **113**, 791–833.

Constant, P., Eybert, M.-C. & Mahéo, R. (1973). Breeding bird populations in conifer plantations in Paimpont Forest, Brittany. (English title.) *Alauda* **41**: 371–384.

Cook, M.J.H. (1982). Breeding status of the Coal Tit. *Scottish Birds* **12**, 97–106.

Cramp, S. (ed.) (1985, 1988, 1992). *The Birds of the Western Palearctic.* Vols. IV–VI. Oxford: Oxford University Press.

Cramp, S. & Perrins, C.M. (eds.) (1993). *The Birds of the Western Palearctic,* Vol. VII. Oxford: Oxford University Press.

Cramp, S. & Simmons, K.E.L. (eds.) (1977, 1979, 1982). *The Birds of the Western Palearctic.* Vols. I–III. Oxford: Oxford University Press.

Cuadrado, M. (1992). Year to year recurrence and site-fidelity of Blackcaps *Sylvia atricapilla* and Robins *Erithacus rubecula* in a Mediterranean wintering area. *Ringing and Migration* **13**, 36–42.

Currie, F.A. & Bamford, R. (1982). The value to birdlife of retaining small conifer stands beyond normal felling age within forests. *Quarterly Journal of Forestry* **76**, 153–160.

Dambach, C.A. (1944). A ten-year ecological study of adjoining grazed and ungrazed woodlands in northeastern Ohio. *Ecological Monographs* **14**, 69–105.

Dhondt, A.A. (1989). Ecological and evolutionary effects of interspecific competition in tits. *Wilson Bulletin* **10**, 198–216.

Dickson, J.G., Conner, R.N. & Williamson, J.H. (1983). Snag retention increases bird use of a clear-cut. *Journal of Wildlife Management* **47**, 799–804.

Dolman, M. & Land, R. (1994). Lowland heathland. In *Habitat Management for Conservation* (ed. W.J. Sutherland & D.A. Hill). Cambridge: Cambridge University Press.

Drent, P.J. & Woldendorp, J.W. (1989). Acid rain and eggshells. *Nature* **339**, 431.

Dudley, N. (1992). *Forests in Trouble: a Review of the Status of Temperate Forests Worldwide.* Gland: World Wide Fund for Nature.

Edgar, R.D.M. (1986). Some results of the study by ringing of warbler migration at Beachy Head from 1960 to 1985. *Sussex Bird Report for 1985* **38**, 76–84.

Edington, J.M. & Edington, M.A. (1972). Spatial patterns and habitat partition in the breeding birds of an upland wood. *Journal of Animal Ecology* **41**, 331–357.

Ekman, J. (1986). Tree use and predator vulnerability of wintering passerines. *Ornis Scandinavica* **17**, 261–267.

Enemar, A., Hanson, S.Å., & Sjöstrand, B. (1965). The composition of the bird fauna in two consecutive breeding seasons in the forests of the Ammarnäs area, Swedish Lapland. *Acta Universitatis Lundensis II*, No. 5.

Enemar, A., & Sjöstrand, B. (1972). Effects of introduction of Pied Flycatchers *Ficedula hypoleuca* on the composition of a passerine bird community. *Ornis Scandinavica* **3**, 79–89.

Enoksson, B. (1990). Autumn territories and population regulation in the Nuthatch *Sitta europaea*: an experimental study. *Journal of Animal Ecology* **59**, 1047–1062.

Enoksson, B. & Nilsson, S.G. (1983). Territory size and population density in relation to food supply in the Nuthatch *Sitta europaea* (Aves). *Journal of Animal Ecology* **52**, 927–935.

Evans, J. (1984). *Silviculture of Broadleaved Woodland*. Forestry Commission Bulletin 62. London: Her Majesty's Stationery Office.

Faliński, J.B. (1986). *Vegetation Dynamics in Temperate Lowland Primeval Forests: Ecological Studies in Białowieża Forest, N.E. Poland*. The Hague: Junk.

Ferry, C. & Frochot, B. (1970). L'avifaune nidificatrice d'une forêt de chênes pédonculés en Bourgogne: étude de deux successions écologiques. *La Terre et la Vie* **2**, 153–250.

Ferry, C. & Frochot, B. (1990). Bird communities of the forests of Burgundy and the Jura (eastern France). In *Biogeography and Ecology of Forest Bird Communities* (ed. A. Keast), pp. 183–195. The Hague: SPB Academic Publishing.

Flousek, J. (1989). Impact of industrial emissions on bird populations breeding in mountain spruce forests in central Europe. *Annales Zoologici Fennici* **26**, 255–263.

Foppen, R. & Reijnen, R. (1994). The effects of car traffic on breeding bird populations in woodland. II. Breeding dispersal of male willow warblers (*Phylloscopus trochilus*) in relation to the proximity of a highway. *Journal of Applied Ecology* **31**, 95–101.

Ford, H.A. (1987). Bird communities on habitat islands in England. *Bird Study* **34**, 205–218.

Forestry Commission (1977–1991). *Forestry Facts and Figures*. (Published annually). Edinburgh: Forestry Commission.

Forestry Commission (1988). *Forests and Water Guidelines*. Edinburgh: Forestry Commission.

Forestry Commission (1989a). *Broadleaves Policy: Progress 1985–88*. Edinburgh: Forestry Commission.

Forestry Commission (1989b). *Forest Landscape Design Guidelines*. Edinburgh: Forestry Commission.

Forestry Commission (1990). *Forest Nature Conservation Guidelines*. London: Her Majesty's Stationery Office.

Foster, D.R. (1988). Disturbance history, community organization and vegetation dynamics of the old-growth Pisgah Forest, south-western New Hampshire, USA. *Journal of Ecology* **76**, 105–134.

French, D.D., Jenkins, D. & Conroy, J.W.H. (1986). Guidelines for managing woods in Aberdeenshire for song birds. In *Trees and Wildlife in the Scottish Uplands* (ed. D. Jenkins), pp. 129–143. ITE Symposium No. 17. Huntingdon: Institute of Terrestrial Ecology.

Frochot, B. (1971). L'evolution saisonniere de l'avifaune dans une futaie de chênes en Bourgogne. *La Terre et la Vie* **188**, 145–182.

Fuller, R.J. (1982). *Bird Habitats in Britain*. Calton: Poyser.

Fuller, R.J. (1987). Composition and Structure of Bird Communities in Britain. Ph.D. thesis, University of London.

Fuller, R.J. (1990). Responses of birds to lowland woodland management in Britain: opportunities for integrating conservation with forestry. *Sitta* **4**, 39–50.

Fuller, R.J. (1991). Effects of woodland edges on songbirds. In *Edge Management in Woodlands* (ed. R. Ferris-Kaan), pp. 31–34. Forestry Commission Occasional Paper 28. Edinburgh: Forestry Commission.

Fuller, R.J. (1992). Effects of coppice management on woodland breeding birds. In *Ecology and Management of Coppice Woodlands* (ed. G.P. Buckley), pp. 169–192. London: Chapman & Hall.

Fuller, R.J. (1994). Relating birds to vegetation: influences of scale, floristics and habitat structure. In *Bird Numbers 1992*. Proceedings 12th International Conference of IBCC and EOAC, The Netherlands. (In the press.)

Fuller, R.J. & Crick, H.Q.P. (1992). Broad-scale patterns in geographical and habitat distribution of migrant and resident passerines in Britain and Ireland. *Ibis* **134** (suppl. 1), 14–20.

Fuller, R.J., Gough S.J. & Marchant, J.H. (1995). Bird populations in new lowland woods: landscape, design and management perspectives. In *The Ecology of Woodland Creation* (ed. R. Ferris-Kaan). London: Wiley. (In the press.)

Fuller, R.J. & Henderson, A.C.B. (1992). Distribution of breeding songbirds in Bradfield Woods, Suffolk, in relation to vegetation and coppice management. *Bird Study* **39**, 73–88.

Fuller, R.J. & Peterken, G.F. (1994). Woodland and scrub. In *Habitat Management for Conservation* (ed. W.J. Sutherland & D.A. Hill). Cambridge: Cambridge University Press.

Fuller, R.J. & Warren, M.S. (1991). Conservation management in ancient and modern woodlands: responses of fauna to edges and rotations. In *The Scientific Management of Temperate Communities for Conservation* (ed. I.F. Spellerberg, F.B. Goldsmith & M.G. Morris), pp. 445–471. British Ecological Society Symposium No. 31. Oxford: Blackwell Scientific Publications.

Fuller, R.J. & Warren, M.S. (1993). *Coppiced Woodlands: their Management for Wildlife*, 2nd edn. Peterborough: Joint Nature Conservation Committee.

Fuller, R.J. & Whittington, P.A. (1987). Breeding bird distribution within Lincolnshire ash-lime woodlands: the influence of rides and the woodland edge. *Acta Oecologica/Oecologica Generalis* **8**, 259–268.

Garcia, E.F.J. (1983). An experimental test of competition for space between Blackcaps *Sylvia atricapilla* and Garden Warblers *Sylvia borin* in the breeding season. *Journal of Animal Ecology* **52**, 795–805.

Gee, A.S. & Stoner, J.H. (1988). The effects of afforestation and acid deposition on the water quality and ecology of upland Wales. In *Ecological Change in the Uplands* (ed. M.B. Usher & D.B.A. Thompson), pp. 273–287. Oxford: Blackwell Scientific Publications.

Gibb, J. (1954). Feeding ecology of tits, with notes on Treecreeper and Goldcrest. *Ibis* **96**, 513–543.

Gibb, J.A. (1960). Populations of tits and Goldcrests and their food supply in pine plantations. *Ibis* **102**, 163–208.

Gibbons, D.W. & Gates, S. (1994). Hypothesis testing with ornithological atlas data: two case studies. In *Bird Numbers 1992*. Proceedings 12th International Conference of IBCC and EOAC, The Netherlands. (In the press.)

Gibbons, D.W., Reid, J.B. & Chapman, R.A. (1993). *The New Atlas of Breeding Birds in Britain and Ireland: 1988–1991*. London: Poyser.

Gibbs, R.G. & Wiggington, M.J. (1973). A breeding bird census in a sessile oakwood at Aber, Caernarvonshire. *Nature in Wales* **13**, 158–162.

Gjerde, I., & Wegge, P. (1989). Spacing pattern, habitat use and survival of Capercaillie in a fragmented winter habitat. *Ornis Scandinavica* **20**, 219–225.

Glas, P. (1960). Factors governing density in the Chaffinch (*Fringilla coelebs*) in different types of wood. *Archives Néerlandaises de Zoologie* **13**, 466–472.

Glue, D. & Morgan, R. (1972). Cuckoo hosts in British habitats. *Bird Study* **19**, 187–192.

Glutz von Blotzheim, U.N. & Bauer, K.M. (1988). *Handbuch der Vögel Mitteleuropas*, Vol. 11, Part 2. Wiesbaden: AULA-Verlag.

Godwin, H. (1975). *The History of the British Flora*, 2nd ed. Cambridge: Cambridge University Press.

Goodier, R. & Bunce, R.G.H. (1977). The native pinewoods of Scotland: the current state of the resource. In *Native Pinewoods of Scotland* (ed. R.G.H. Bunce & J.N.R. Jeffers), pp. 78–87. Cambridge: Institute of Terrestrial Ecology.

Goodwin, D. (1976). *Crows of the World*. London: British Museum (Natural History).

Gosler, A.M. (1990). The birds of Wytham – an historical survey. *Fritillary* **1**, 29–74.

Granval, Ph. & Muys, B. (1992). Management of forest soils and earthworms to improve Woodcock (*Scolopax* sp.) habitats: a literature survey. *Gibier Faune Sauvage* **9**, 243–255.

Greenwood, J.J.D. & Baillie, S.R. (1991). Effects of density-dependence and weather on population changes of English passerines using a non-experimental paradigm. *Ibis* **133** (suppl. 1), 121–133.

Greenwood, P.J., Harvey, P.H. & Perrins, C.M. (1979). The role of dispersal in the Great Tit (*Parus major*): the causes, consequences and heritability of natal dispersal. *Journal of Animal Ecology* **48**, 123–142.

Gunnarsson, B. (1990). Vegetation structure and the abundance and size distribution of spruce-living spiders. *Journal of Animal Ecology* **59**, 743–752.

Gustafsson, L. (1987). Interspecific competition lowers fitness in Collared Flycatchers *Ficedula albicollis*: an experimental demonstration. *Ecology* **68**, 291–296.

Haapanen, A. (1966). Bird fauna of the Finnish forests in relation to forest succession II. *Annales Zoologici Fennici* **3**, 176–200.

Haemig, P.D. (1992). Competition between ants and birds in a Swedish forest. *Oikos* **65**, 479–483.

Hågvar, S., Hågvar, G. & Mønness, E. (1990). Nest site selection in Norwegian woodpeckers. *Holarctic Ecology* **13**, 156–165.

Haila, Y., Hanski, I.K. & Raivio, S. (1987). Breeding bird distribution in fragmented coniferous taiga, southern Finland. *Ornis Fennica* **64**, 90–106.

Haila, Y. & Järvinen, O. (1990). Northern conifer forests and their bird species assemblages. In *Biogeography and Ecology of Forest Bird Communities* (ed. A. Keast), pp. 61–85. The Hague: SPB Academic Publishing.

Hake, M. (1991). The effects of needle loss in coniferous forests in south-west Sweden on the winter foraging behaviour of willow tits *Parus montanus*. *Biological Conservation* **58**, 357–366.

Hansson, L. (1983). Bird numbers across edges between mature conifer forest and clearcuts in central Sweden. *Ornis Scandinavica* **14**, 97–103.

Harding, P.T. & Rose, F. (1986). *Pasture-woodlands in Lowland Britain: a Review of their Importance for Wildlife Conservation.* Huntingdon: Institute of Terrestrial Ecology.

Harris, E. & Harris, J. (1991). *Wildlife Conservation in Managed Woodlands and Forests.* Oxford: Basil Blackwell.

Harris, M.J. & Kent, M. (1987). Ecological benefits of the Bradford-Hutt system of commercial forestry I. Ground flora and the light climate. *Quarterly Journal of Forestry* **81**, 145–157.

Harrison, C. (1975). *The Nests, Eggs and Nestlings of British and European Birds.* London: Collins.

Harrison, C. (1982). *An Atlas of the Birds of the Western Palaearctic.* London: Collins.

Harrison, C.J. (1988). *The History of the Birds of Britain.* London: Collins.

Hartley, P.H.T. (1987). Ecological aspects of the foraging behaviour of Crested Tits *Parus cristatus. Bird Study* **34**, 107–111.

Helle, P. & Fuller, R.J. (1988). Migrant passerine birds in European forest successions in relation to vegetation height and geographical position. *Journal of Animal Ecology* **57**, 565–580.

Helle, P. & Järvinen, O. (1986). Population trends of North Finnish land birds in relation to their habitat selection and changes in forest structure. *Oikos* **46**, 107–115.

Helle, P. & Mönkkönen, M. (1990). Forest successions and bird communities: theoretical aspects and practical implications. In *Biogeography and Ecology of Forest Bird Communities* (ed. A. Keast), pp. 299–318. The Hague: SPB Academic Publishing.

Henderson, I.G. (1989). The exploitation of tits *Parus* species, Long-tailed Tits *Aegithalos caudatus* and Goldcrests *Regulus regulus* by Treecreepers *Certhia familiaris*: a behavioural study. *Bird Study* **36**, 99–104.

Henttonen, H. (1989). Does an increase in the rodent and predator densities, resulting from modern forestry, contribute to the long-term decline in Finnish tetraonids? (English title.) *Suomen Riista* **35**, 83–90.

Herrera, C.M. (1988). Annual variation in the abundance of frugivorous birds and its relationship to the availability of fruits. (English title.) *Ardeola* **35**, 135–142.

Herremans, M. (1993). Clustering of territories in the Wood Warbler *Phylloscopus sibilatrix. Bird Study* **40**, 12–23.

Hibberd, B.G. (1985). Restructuring of plantations in Kielder Forest District. *Forestry* **58**, 119–129.

Hibberd, B.G. (ed.) (1989). *Urban Forestry Practice.* Forestry Commission Handbook 5. London: Her Majesty's Stationery Office.

Hill, D.A., Lambton, S., Proctor, I. & Bullock, I. (1991). Winter bird communities in woodland in the Forest of Dean, England, and some implications of livestock grazing. *Bird Study* **38**, 57–70.

Hill, D., Taylor, S., Thaxton, R., Amphlet, A. & Horn, W. (1990). Breeding bird communities of native pine forest, Scotland. *Bird Study* **37**, 133–141.

Hinsley, S., Bellamy, P. & Newton, I. (1992). Habitat fragmentation, landscape ecology and birds. *Institute of Terrestrial Ecology Report 1991–1992*, 19–21.

Hirons, G. (1982). The diet and behaviour of woodcock (*Scolopax rusticola* L.) in winter. *Transactions International Congress of Game Biology* **14**, 233–235.

Hirons, G. & Johnson, T.H. (1987). A quantitative analysis of habitat preferences of Woodcock *Scolopax rusticola* in the breeding season. *Ibis* **129**, 371–381.

Hoelzel, A.R. (1989). Territorial behaviour of the Robin *Erithacus rubecula*: the importance of vegetation density. *Ibis* **131**, 432–436.

Hogstad, O. (1971). Stratification in winter feeding of the Great Spotted Woodpecker *Dendrocopos major* and the Three-toed Woodpecker *Picoides tridactylus*. *Ornis Scandinavica* **2**, 143–146.

Holmes, R.T. (1990). The structure of a temperate deciduous forest bird community: variability in time and space. In *Biogeography and Ecology of Forest Bird Communities* (ed. A. Keast), pp. 121–139. The Hague: SPB Academic Publishing.

Holmes, R.T. & Robinson, S.K. (1981). Tree species preferences of foraging insectivorous birds in a northern hardwoods forest. *Oecologia* **48**, 31–35.

Holmes, R.T., Schultz, J.C. & Nothnagle, P. (1979). Bird predation on forest insects: an exclosure experiment. *Science* **206**, 462–463.

Hölzinger, J. & Kroymann, B. (1984). Consequences of the 'Waldsterben' on the avifauna in southwest Germany. (English title.) *Ökologie der Vögel* **6**, 203–212.

Hope Jones, P. (1972). Succession in breeding bird populations of sample Welsh oakwoods. *British Birds* **65**, 291–299.

Hope Jones, P. (1975). Winter bird populations in a Merioneth oakwood. *Bird Study* **22**, 25–34.

Hughes, S.W.M. & Griffiths, A.J. (1983). The birds of Iping Common. *Sussex Bird Report for 1982* **35**, 88–94.

Humphrey, J. (1992). The Natural Regeneration of Scottish Oakwoods. Ph.D. thesis, University of Aberdeen.

Hunt, J. (1990). Abernethy Forest. *Scottish Bird News* **19**, 6–7.

Huntley, B. & Birks, H.J.B. (1983). *An Atlas of Past and Present Pollen Maps for Europe: 0–13,000 Years Ago*. Cambridge: Cambridge University Press.

Hutchinson, C.D. (1989). *Birds in Ireland*. Calton: Poyser.

Intergovernmental Panel on Climate Change (1990). *Climate Change: the IPCC Scientific Assessment*. (ed. J.T. Houghton, G.J. Jenkins & J.J. Ephraums). Cambridge: Cambridge University Press.

Irvine, J. (1977). Breeding birds in New Forest broad-leaved woodland. *Bird Study* **24**, 105–111.

Jahn, G. (1991). Temperate deciduous forests of Europe. In *Temperate Deciduous Forests* (ed. E. Röhrig & B. Ulrich), pp. 377–502. Amsterdam: Elsevier.

James, F.C. & Rathbun, S. (1981). Rarefaction, relative abundance, and diversity of avian communities. *Auk* **98**, 785–800.

James, F.C. & Wamer, N.O. (1982). Relationships between temperate forest bird communities and vegetation structure. *Ecology* **63**, 159–171.

Jansen, P.B. & de Nie, H.W. (1986). Thirty years of passerine breeding bird monitoring in a mixed wood (English title). *Limosa* **59**, 127–134.

Jenni, L. (1986). The importance of large roosts of Bramblings (*Fringilla montifringilla*) in beech-mast areas. (English title.) *Ornithologische Beobachter* **83**, 267–268.

Joensen, A.H. (1965). An investigation on bird populations in four deciduous forest areas on Als in 1962 and 1963. (English title.) *Dansk Ornithologisk Forenings Tidsskrift* **59**, 115–186.

Kavanagh, B. (1990). Bird communities of two short rotation forestry plantations on cutover peatland. *Irish Birds* **4**, 169–180.

Kelly, J.P. (1993). The effect of nest predation on habitat selection by Dusky Flycatchers in limber pine-juniper woodland. *Condor* **95**, 83–93.

Kemp, J.B. (1982). Winter roosts and habitats of Long-eared and Short-eared Owls. *British Birds* **75**, 334–335.

Kirby, K.J. & Heap, J.R. (1984). Forestry and nature conservation in Romania. *Quarterly Journal of Forestry* **78**, 145–155.

Kirby, K.J., Webster, S.D. & Antczak, A. (1991). Effects of forest management on stand structure and the quantity of fallen dead wood: some British and Polish examples. *Forest Ecology and Management* **43**, 167–174.

Kirby, P. (1992). *Habitat Management for Invertebrates: a Practical Handbook.* Sandy: Royal Society for the Protection of Birds.

Krause, G.H.M. (1989). Forest decline in central Europe: the unravelling of multiple causes. In *Toward a More Exact Ecology* (ed. P.J. Grubb & J.B. Whittaker), pp. 377–399. Oxford: Blackwell Scientific Publications.

Krebs, J.R. (1971). Territory and breeding density in the Great Tit, *Parus major* L. *Ecology* **52**, 1–22.

Lack, D. (1971). *Ecological Isolation in Birds.* Oxford: Blackwell Scientific Publications.

Lack, D. & Lack, E. (1951). Further changes in bird-life caused by afforestation. *Journal of Animal Ecology* **20**, 173–179.

Lack, D. & Venables, L.S.V. (1939). The habitat distribution of British woodland birds. *Journal of Animal Ecology* **8**, 39–71.

Lack, P. (1986). *The Atlas of Wintering Birds in Britain and Ireland.* Calton: Poyser.

Last, F.T., Jeffers, J.N.R., Bunce, R.G.H., Claridge, C.J., Baldwin, M.B. & Cameron, R.J. (1986). Whither forestry? The scene in AD 2025. In *Trees and Wildlife in the Scottish Uplands* (ed. D. Jenkins), pp. 20–32. ITE Symposium No. 17. Huntingdon: Institute of Terrestrial Ecology.

Lawn, M.R. (1982). Pairing systems and site tenacity of the Willow Warbler *Phylloscopus trochilus* in southern England. *Ornis Scandinavica* **13**, 193–199.

Lawn, M.R. (1984). Premigratory dispersal of juvenile Willow Warblers *Phylloscopus trochilus* in southern England. *Ringing and Migration* **5**, 125–131.

Lebreton, P. & Broyer, J. (1981). Contribution à l'étude des relations avifaune/altitiude. I Au niveau de la région Rhône-Alpes. *L'Oiseau et la Revue Française d'Ornithologie* **51**, 265–285.

Leisler, B. & Thaler, E. (1982). Differences in morphology and foraging behaviour in the goldcrest *Regulus regulus* and firecrest *R. ignicapillus*. *Annales Zoologici Fennici* **19**, 277–284.

Leverton, R. (1986). Passage and wintering thrushes at a downland site. *Sussex Bird Report for 1985* **38**, 67–75.

Lindroth, R.L., Kinney, K.K. & Platz, C.L. (1993). Responses of deciduous trees to elevated atmospheric CO_2: productivity, phytochemistry, and insect performance. *Ecology* **74**, 763–777.

Ling, K.A., Power, S.A. & Ashmore, M.R. (1993). A survey of the health of *Fagus sylvatica* in southern Britain. *Journal of Applied Ecology* **30**, 295–306.

Little, B. & Davison, M. (1992). Merlins *Falco columbarius* using crow nests in Kielder Forest, Northumberland. *Bird Study* **39**, 13–16.

Locke, G.M.L. (1987). *Census of Woodlands and Trees 1979–82.* Forestry Commission Bulletin 63. London: Her Majesty's Stationery Office.

Lundberg, A. & Alatalo, R.V. (1992). *The Pied Flycatcher.* London: Poyser.

Lynch, J.F. & Whigham, D.F. (1984). Effects of forest fragmentation on breeding bird communities in Maryland, USA. *Biological Conservation* **28**, 287–324.

Mabey, R. (1993). *Whistling in the Dark: in Pursuit of the Nightingale.* London: Sinclair-Stevenson.

MacArthur, R.H. & MacArthur, J. (1961). On bird species diversity. *Ecology* **42**, 594–598.

MacArthur, R. & Wilson, E.O. (1967). *The Theory of Island Biogeography*. Princeton: Princeton University Press.

Mackenzie, J.M.D. (1945). The preference shown by birds for different species of trees in plantations. *Forestry* **19**, 97–112.

Magrath, R.D. (1991). Nestling weight and juvenile survival in the Blackbird, *Turdus merula*. *Journal of Animal Ecology* **60**, 335–351.

Mannan, R.W. & Meslow, E.C. (1984). Bird populations and vegetation characteristics in managed and old-growth forests, north-eastern Oregon. *Journal of Wildlife Management* **48**, 1219–1238.

Marchant, J.H., Hudson, R., Carter, S.P. & Whittington, P. (1990). *Population Trends in British Breeding Birds*. Tring: British Trust for Ornithology.

Marquiss, M. (1989). Grey Herons *Ardea cinerea* breeding in Scotland: numbers, distribution, and census techniques. *Bird Study* **36**, 181–191.

Marquiss, M. & Newton, I. (1982). Habitat preference in male and female Sparrowhawks. *Ibis* **124**, 324–328.

Marrs, R.H. (1985). The use of Krenite to control birch on lowland heather. *Biological Conservation* **32**, 149–164.

Marrs, R.H., Hicks, M.J. & Fuller, R.M. (1986). Losses of lowland heath through succession at four sites in Breckland, East Anglia, England. *Biological Conservation* **36**, 19–38. (Elsevier Science Ltd., Kidlington, UK.)

Martin, T.E. & Karr, J.R. (1986). Patch utilization by migrating birds: resource oriented? *Ornis Scandinavica* **17**, 165–174.

Matthews, J.D. (1989). *Silvicultural Systems*. Oxford: Oxford University Press.

Matthysen, E. (1990). Behavioural and ecological correlates of territory quality in the Eurasian Nuthatch (*Sitta europaea*). *Auk* **107**, 86–95.

McClanahan, T.R. & Wolfe, R.W. (1993). Accelerating forest succession in a fragmented landscape: the role of birds and perches. *Conservation Biology* **7**, 279–288.

McCleery, R.H. & Perrins, C.M. (1985). Territory size, reproductive success and population dynamics in the Great Tit *Parus major*. In *Behavioural Ecology: Ecological Consequences of Adaptive Behaviour* (ed. R.M. Sibly & R.H. Smith), pp. 353–373. Oxford: Blackwell Scientific Publications.

McIntosh, R. (1989). Forest design: Kielder Forest restructuring. *Timber Grower* **Autumn**, 19–20.

Melluish, W.D. (1969). Further notes on the bird population of grassland with encroaching scrub at Bookham Common. *London Naturalist* **48**, 135–142.

Mitchell, F.J.G. & Kirby, K.J. (1990). The impact of large herbivores on the conservation of semi-natural woods in the British uplands. *Forestry* **63**, 333–353.

Mitchell, P.L. & Kirby, K.J. (1989). *Ecological Effects of Forestry Practices in Long-established Woodland and their Implications for Nature Conservation*. Oxford Forestry Institute Occasional Paper No. 39. Oxford: Oxford Forestry Institute.

Mock, D., Lamey, T.C. & Thompson, D.B.A. (1988). Falsifiability and the information centre hypothesis. *Ornis Scandinavica* **19**, 231–248.

Möckel, R. (1992). Effect of 'Waldsterben' (forest damage due to airborne pollution) on the population dynamics of Coal Tit (*Parus ater*) and Crested Tit (*Parus cristatus*) in the western Erzgebirge. (English title.) *Ökologie der Vögel* **14**, 1–100.

Mönkkönen, M. & Helle, P. (1989). Migratory habits of birds breeding in different stages of forest succession: a comparison between the Palaearctic and the Nearctic. *Annales Zoologici Fennici* **26**, 323–330.

Moore, N.W. & Hooper, M.D. (1975). On the number of bird species in British woods. *Biological Conservation* **8**, 239–250.

Moreau, R.E. (1954). The main vicissitudes of the European avifauna since the Pliocene. *Ibis* **96**, 411–431.

Morgan, R.A. (1978). Changes in the breeding bird community at Gibraltar Point, Lincolnshire, between 1965 and 1974. *Bird Study* **25**, 51–58.

Morris, A., Burges, D., Fuller, R.J., Evans, A.D. & Smith, K.W. (1995). The status and distribution of Nightjars *Caprimulgus europaeus* in Britain in 1992. *Bird Study* (in the press).

Morrison, M.L., Timossi, I.C., With, K.A. & Manley, P.N. (1985). Use of tree species by forest birds during winter and summer. *Journal of Wildlife Management* **49**, 1098–1102.

Moss, D. (1978). Diversity of woodland song-bird populations. *Journal of Animal Ecology* **47**, 521–527.

Moss, D., Taylor, P.N. & Easterbee, N. (1979). The effects on song-bird populations of upland afforestation with spruce. *Forestry* **52**, 129–147.

Moss, R. (1986). Rain, breeding success and distribution of Capercaillie *Tetrao urogallus* and Black Grouse *Tetrao tetrix* in Scotland. *Ibis* **128**, 65–72.

Moss, R. & Picozzi, N. (1994). *Management of Forests for Capercaillie in Scotland.* London: Her Majesty's Stationery Office (in the press).

Murton, R.K. & Westwood, N.J. (1974). Some effects of agricultural change on the English avifauna. *British Birds* **67**, 41–69.

Newton, I. (1972). *Finches.* London: Collins.

Newton, I. (1986a). *The Sparrowhawk.* Calton: Poyser.

Newton, I. (1986b). Principles underlying bird numbers in Scottish woodlands. In *Trees and Wildlife in the Scottish Uplands* (ed. D. Jenkins), pp. 121–128. ITE Symposium No. 17. Huntingdon: Institute of Terrestrial Ecology.

Newton, I. (1991). Habitat variation and population regulation in Sparrowhawks. *Ibis* **133** (suppl. 1), 76–88.

Newton, I. & Moss, D. (1977). Breeding birds of Scottish pinewoods. In *Native Pinewoods of Scotland.* (ed. R.G.H. Bunce & J.N.R. Jeffers), pp. 26–34. Cambridge: Institute of Terrestrial Ecology.

Newton, I., Wyllie, I. & Mearns, R. (1986). Spacing of Sparrowhawks in relation to food supply. *Journal of Animal Ecology* **55**, 361–370.

Nilsson, S.G. (1976). Habitat, territory size, and reproductive success in the Nuthatch *Sitta europaea*. *Ornis Scandinavica* **7**, 179–184.

Nilsson, S.G. (1979a). Seed density, cover, predation and the distribution of birds in a beech wood in southern Sweden. *Ibis* **121**, 177–185.

Nilsson, S.G. (1979b). Density and species richness of some forest bird communities in south Sweden. *Oikos* **33**, 392–401.

Nilsson, S.G. (1979c). Effect of forest management on the breeding bird community in southern Sweden. *Biological Conservation* **16**, 135–143.

Nilsson, S.G. (1984). The evolution of nest-site selection among hole-nesting birds: the importance of nest predation and competition. *Ornis Scandinavica* **15**, 167–175.

Nilsson, S.G. (1985). Ecological and evolutionary interactions between reproduction of beech *Fagus sylvatica* and seed eating animals. *Oikos* **44**, 157–164.

Nilsson, S.G. (1986a). Evolution of hole-nesting in birds: on balancing selection pressures. *Auk* **103**, 432–435.

Nilsson, S.G. (1986b). Are bird communities in small biotope patches random samples from communities in large patches? *Biological Conservation* **38**, 179–204.

Nilsson, S.G. (1987). Limitation and regulation of population density in the Nuthatch *Sitta europaea* (Aves) breeding in natural cavities. *Journal of Animal Ecology* **56**, 921–937.

Nilsson, S.G., Björkman, C., Forslund, P. & Högland, J. (1985). Nesting holes and food supply in relation to forest bird densities on islands and mainland. *Oecologia* **66**, 516–521.

O'Connor, R.J. (1985). Behavioural regulation of bird populations: a review of habitat use in relation to migration and residency. In *Behavioural Ecology: Ecological Consequences of Adaptive Behaviour* (ed. R.M. Sibly & R.H. Smith), pp. 105–142. Oxford: Blackwell Scientific Publications.

O'Connor, R.J. & Shrubb, M. (1986). *Farming and Birds*. Cambridge: Cambridge University Press.

Olsson, O., Nilsson, I.N., Nilsson, S.G., Pettersson, B., Stagen, A. & Wiktander, U. (1992). Habitat preferences of the Lesser Spotted Woodpecker *Dendrocopos minor*. *Ornis Fennica* **69**, 119–125.

Opdam, P., Rijsdijk, G. & Hustings, F. (1985). Bird communities in small woods in an agricultural landscape: effects of area and isolation. *Biological Conservation* **34**, 333–352.

Osborne, P. (1983). The influence of Dutch elm disease on bird population trends. *Bird Study* **30**, 27–38.

Osborne, P. (1985). Some effects of Dutch elm disease on the birds of a Dorset dairy farm. *Journal of Applied Ecology* **22**, 681–691.

Parr, S.J. (1991). Occupation of new conifer plantations by Merlins in Wales. *Bird Study* **38**, 103–111.

Parr, R. & Watson, A. (1988). Habitat preferences of Black Grouse on moorland-dominated ground in north-east Scotland. *Ardea* **76**, 175–180.

Peck, K.M. (1989). Tree species preferences shown by foraging birds in forest plantations in Northern England. *Biological Conservation* **48**, 41–57.

Perrins, C.M. (1979). *British Tits*. London: Collins.

Peterken, G.F. (1991). Ecological issues in the management of woodland nature reserves. In *The Scientific Management of Temperate Communities for Conservation* (ed. I.F. Spellerberg, F.B. Goldsmith & M.G. Morris), pp. 245–272. British Ecological Society Symposium No. 31. Oxford: Blackwell Scientific Publications.

Peterken, G.F. (1992). Conservation of old growth: a European perspective. *Natural Areas Journal* **12**, 10–19.

Peterken, G.F. (1993). *Woodland Conservation and Management*, 2nd edn. London: Chapman & Hall.

Peterken, G.F. & Allison, H. (1989). *Woods, Trees and Hedges: a Review of Changes in the British Countryside*. Focus on Nature Conservation 22. Peterborough: Nature Conservancy Council.

Peterken, G.F., Ausherman, D., Buchenau, M. & Forman, R.T.T. (1992). Old-growth conservation within British upland conifer plantations. *Forestry* **65**, 127–144.

Petty, S.J. & Anderson, D. (1986). Breeding by Hen Harriers *Circus cyaneus* on restocked sites in upland forests. *Bird Study* **33**, 177–178.

Petty, S.J. & Avery, M.I. (1990). *Forest Bird Communities*. Forestry Commission Occasional Paper 26. Edinburgh: Forestry Commission.

Picozzi, N., Catt, D.C. & Moss, R. (1992). Evaluation of Capercaillie habitat. *Journal of Applied Ecology* **29**, 751–762.

Pigott, C.D. (1983). Regeneration of oak-birch woodland following exclusion of sheep. *Journal of Ecology* **71**, 629–646.

Polunin, O. & Walters, M. (1985). *A Guide to the Vegetation of Britain and Europe.* Oxford: Oxford University Press.

Prinz, B. (1987). Causes of forest damage in Europe. *Environment* **29**, 11–37.

Putman, R.J. (1986). *Grazing in Temperate Ecosystems: Large Herbivores and the Ecology of the New Forest.* Beckenham: Croom Helm.

Putman, R.J., Edwards, P.J., Mann, J.C.E., How, R.C. & Hill, S.D. (1989). Vegetational and faunal changes in an area of heavily grazed woodland following relief of grazing. *Biological Conservation* **47**, 13–32.

Rackham, O. (1980). *Ancient Woodland: its History, Vegetation and Uses in England.* London: Arnold.

Rackham, O. (1986). *The History of the Countryside.* London: Dent.

Rackham, O. (1990). *Trees and Woodland in the British Landscape,* 2nd edn. London: Dent.

Ratcliffe, D. (1990). *Bird Life of Mountain and Upland.* Cambridge: Cambridge University Press.

Ratcliffe, P.R. & Petty, S.J. (1986). The management of commercial forests for wildlife. In *Trees and Wildlife in the Scottish Uplands* (ed. D. Jenkins), pp. 177–187. ITE Symposium 17. Huntingdon: Institute of Terrestrial Ecology.

Ravenscroft, N.O.M. (1989). The status and habitat of the Nightjar *Caprimulgus europaeus* in coastal Suffolk. *Bird Study* **36**, 161–169.

Reed, J.M. & Dobson, A.P. (1993). Behavioural constraints and conservation biology: conspecific attraction and recruitment. *Trends in Ecology and Evolution* **8**, 253–255.

Reed, T.M. (1982). Interspecific territoriality in the Chaffinch and Great Tit on islands and the mainland of Scotland: playback and removal experiments. *Animal Behaviour* **30**, 171–181.

Reijnen, M.J.S.M. & Thissen, J.B.M. (1987). Effects from road traffic on breeding-bird populations in woodland. *Annual Report 1986,* pp. 121–132. Leersum: Research Institute for Nature Management.

Reijnen, R. & Foppen, R. (1994). The effects of car traffic on breeding bird populations in woodland. I. Evidence of reduced habitat quality for willow warblers (*Phylloscopus trochilus*) breeding close to a highway. *Journal of Applied Ecology* **31**, 85–94.

Reinikainen, A. (1937). The irregular migrations of the Crossbill *Luxia c. curvirostra,* and their relation to the cone-crop of the conifers. *Ornis Fennica* **14**, 55–64.

Roberts, A.J., Russell, C., Walker, G.J. & Kirby, K.J. (1992). Regional variation in the origin, extent and composition of Scottish woodland. *Botanical Journal of Scotland* **46**, 167–189.

Roberts, T.M., Skeffington, R.A. & Blank, L.W. (1989). Causes of Type 1 spruce decline in Europe. *Forestry* **62**, 179–222.

Robertson, P.A. (1992). *Woodland Management for Pheasants.* Forestry Commission Bulletin 106. London: Her Majesty's Stationery Office.

Robertson, P.A., Woodburn, M.I.A. & Hill, D.A. (1993). Factors affecting winter pheasant density in British woodlands. *Journal of Applied Ecology* **30**, 459–464.

Rodwell, J.S. (1991). *British Plant Communities. Vol. 1. Woodlands and Scrub.* Cambridge: Cambridge University Press.

Rolstad, J. & Wegge, P. (1987). Distribution and size of Capercaillie leks in relation to old forest fragmentation. *Oecologia* **72**, 389–394.

Root, T.L. & Schneider, S.H. (1993). Can large-scale climatic models be linked with multiscale ecological studies? *Conservation Biology* **7**, 256–270.

Rudnicky, T.C. & Hunter, M.L. (1993). Avian nest predation in clearcuts, forests and edges in a forest-dominated landscape. *Journal of Wildlife Management* **57**, 358–364.

Saether, B.-E. (1983). Habitat selection, foraging niches and horizontal spacing of Willow Warbler *Phylloscopus trochilus* and Chiffchaff *P. collybita* in an area of sympatry. *Ibis* **125**, 24–32.

Schmidt, K.-H., März, M. & Matthysen, E. (1992). Breeding success and laying date of Nuthatches *Sitta europaea* in relation to habitat, weather and breeding density. *Bird Study* **39**, 23–30.

Scott, D., Clarke, R., & Shawyer, C.R. (1991). Hen Harriers breeding in a tree-nest. *Irish Birds* **4**, 413–417.

Sharrock. J.T.R. (1976). *The Atlas of Breeding Birds in Britain and Ireland.* Berkhamsted: Poyser.

Shaw, G. & Livingstone, J. (1991). Goldfinches and other birds eating Sitka spruce seed. *BTO News* **174**, 8–9.

Simms, E. (1971). *Woodland Birds.* London: Collins.

Sinclair, A.R.E. (1989). Population regulation in animals. In *Ecological Concepts* (ed. J.M. Cherrett), pp. 197–241. Oxford: Blackwell Scientific Publications.

Smith, K.W. (1987). The ecology of the Great Spotted Woodpecker. *RSPB Conservation Review* **1**, 74–77.

Smith, K.W. (1988). Breeding bird communities of commercially managed broadleaved plantations. *RSPB Conservation Review* **2**, 43–46.

Smith, K.W. (1992). Bird populations: effects of tree species mixtures. In *The Ecology of Mixed-Species Stands of Trees* (ed. M.G.R. Cannell, D.C. Malcolm & P.A. Robertson), pp. 233–242. Oxford: Blackwell Scientific Publications.

Smith, K.W., Averis, B. & Martin, J. (1985). The breeding bird community of oak plantations in the Forest of Dean, southern England. *Acta Oecologia/ Oecologia Generalis* **8**, 209–217.

Smith, K.W., Burges, D.J. & Parks, R.A. (1992). Breeding bird communities of broadleaved plantation and ancient pasture woodlands of the New Forest. *Bird Study* **39**, 132–141.

Snow, B.K & Snow, D.W. (1984). Long-term defence of fruit by Mistle Thrushes *Turdus viscivorus*. *Ibis* **126**, 39–49.

Snow, B. & Snow, D. (1988). *Birds and Berries.* Calton: Poyser.

Snow, D.W. (1988). *A study of Blackbirds.* London: British Museum (Natural History).

Southern, H.N. (1970). The natural control of a population of Tawny Owls (*Strix aluco*). *Journal of Zoology* **162**, 197–285.

Spencer, J.W. & Kirby, K.J. (1992). An inventory of ancient woodland for England and Wales. *Biological Conservation* **62**, 77–93.

Spray, C.J., Crick, H.Q.P. & Hart, A.D.M. (1987). Effects of aerial applications of fenitrothion on bird populations of a Scottish pine plantation. *Journal of Applied Ecology* **24**, 29–47.

Šťastny, K. & Bejček, V. (1985). Bird communities of spruce forests affected by industrial emissions in the Krušné Hory (Ore Mountains). In *Bird Census and Atlas Studies* (ed. K. Taylor, R.J. Fuller & P.C. Lack), pp. 243–253. Proceedings VIII International Conference on Bird Census and Atlas Work. Tring: British Trust for Ornithology.

Stowe, T.J. (1987). The management of sessile oakwoods for Pied Flycatchers. *RSPB Conservation Review* **1**, 78–83.

Suhonen, J. (1993). Predation risk influences the use of foraging sites by tits. *Ecology* **74**, 1197–1203.

Summers-Smith, J.D. (1988). *The Sparrows: A Study of the Genus Passer*. Calton: Poyser.

Summers-Smith, J.D. (1989). A history of the status of the Tree Sparrow *Passer montanus* in the British Isles. *Bird Study* **36**, 23–31.

Swenson, J.E. (1993). The importance of alder to Hazel Grouse in Fennoscandian boreal forest: evidence from four levels of scale. *Ecography* **16**, 37–46.

Sykes, J.M., Lowe, V.P.W. & Briggs, D.R. (1989). Some effects of afforestation on the flora and fauna of an upland sheepwalk during 12 years after planting. *Journal of Applied Ecology* **26**, 299–320.

Temrin, H., Mallner, Y. & Winder, M. (1984). Observations on polyterritoriality and singing behaviour in the Wood Warbler *Phylloscopus sibilatrix*. *Ornis Scandinavica* **15**, 62–72.

Thomas, J.W., Forsman, E.D., Lint, J.B., Meslow, E.C., Noon, B.R. & Verner, J. (1990). *A Conservation Strategy for the Northern Spotted Owl*. Portland, Oregon: United States Government Printing Office.

Thompson. D.B.A., Stroud, D.A. & Pienkowski, M.W. (1988). Afforestation and upland birds: consequences for population ecology. In *Ecological Change in the Uplands* (ed. M.B. Usher & D.B.A. Thompson), pp. 237–259. Oxford: Blackwell Scientific Publications.

Tjernberg, M. (1983). Habitat and nest site features of Golden Eagle *Aquila chrysaetos* (L.) in Sweden. *Viltrevy* **12**, 131–163.

Tomiałojć, L. & Wesołowski, T. (1990). Bird communities of the primaeval temperate forest of Białowieża, Poland. In *Biogeography and Ecology of Forest Bird Communities* (ed. A. Keast), pp. 141–165. The Hague: SPB Academic Publishing.

Tomiałojć, L., Wesołowski, T. & Walankiewicz, W. (1984). Breeding bird community of a primaeval temperate forest (Białowieża National Park, Poland). *Acta Ornithologica* **20**, 241–310.

Tubbs, C.R. (1986). *The New Forest*. London: Collins.

Tubbs, C.R. & Tubbs, J.M. (1985). Buzzards *Buteo buteo* and land use in the New Forest, Hampshire, England. *Biological Conservation* **31**, 41–65.

Ulfstrand, S. (1975). Bird flocks in relation to vegetation diversification in a south Swedish coniferous plantation during winter. *Oikos* **26**, 65–73.

Väisänen, R.A., Järvinen, O. & Rauhala, P. (1986). How are extensive, human-caused habitat alterations expressed on the scale of local bird populations in boreal forests? *Ornis Scandinavica* **17**, 282–292.

van der Zande, A.N., Berkhuizen, J.C., van Latesteijn, H.C., Keurster, W.J. & Poppelaars, A.J. (1984). Impact of outdoor recreation on the density of a number of breeding bird species in woods adjacent to urban residential areas. *Biological Conservation* **30**, 1–39.

van Dorp, D. & Opdam, P.F.M. (1987). Effects of patch size, isolation and regional abundance on forest bird communities. *Landscape Ecology* **1**, 59–73.

Verner, J. (1992). Data needs for avian conservation biology: have we avoided critical research? *Condor* **94**, 301–303.

Verner, J. & Larson, T.A. (1989). Richness of breeding bird species in mixed-conifer forests of the Sierra Nevada, California. *Auk* **106**, 447–463.

Verstrael, T. (1989). Vijf jaar Broedvogel Monitoring Projeckt in Flevoland. *Grauwe Gans* **5**, 65–92.

Virkkala, R. (1987). Effects of forest management on birds breeding in northern Finland. *Annales Zoologici Fennici* **24**, 281–294.

Virkkala, R. (1990). Ecology of the Siberian Tit *Parus cinctus* in relation to habitat quality: effects of forest management. *Ornis Scandinavica* **21**, 139–146.

Virkkala, R., Heinonen, M. & Routasuo P. (1991). The response of northern taiga birds to storm disturbance in the Koilliskaira National Park, Finnish Lapland. *Ornis Fennica* **68**, 123–126.

Virkkala, R. & Liehu, H. (1990). Habitat selection by the Siberian Tit *Parus cinctus* in virgin and managed forests in northern Finland. *Ornis Fennica* **67**, 1–12.

Voous, K.H. (1960). *Atlas of European Birds*. London: Nelson.

Voous, K.H. (1977). *List of Recent Holarctic Bird Species*. London: British Ornithologists' Union.

von Haartman, L. (1971). Population dynamics. In *Avian Biology 1* (ed. D.S. Farner & J.R. King), pp. 391–459. London: Academic Press.

Walters Davies, P. & Davis, P.E. (1973). The ecology and conservation of the Red Kite in Wales. *British Birds* **66**, 183–224, 241–270.

Warren, M.S. & Fuller, R.J. (1993). *Woodland Rides and Glades: their Management for Wildlife*, 2nd edn. Peterborough: Joint Nature Conservation Committee.

Watson, J. (1979). Food of Merlins nesting in young conifer forest. *Bird Study* **26**, 253–258.

Webb, N.R. (1990). Changes on the heathlands of Dorset, England, between 1978 and 1987. *Biological Conservation* **51**, 273–286.

Wesołowski, T. (1983). The breeding ecology and behaviour of Wrens *Troglodytes troglodytes* under primaeval and secondary conditions. *Ibis* **125**, 499–515.

Wesołowski, T. (1989). Nest-sites of hole-nesters in a primaeval temperate forest (Białowieża National Park, Poland). *Acta Ornithologica* **25**, 321–351.

Widén, P. (1989). The hunting habitats of Goshawks *Accipiter gentilis* in boreal forests of central Sweden. *Ibis* **131**, 205–231.

Wiens, J. (1989). *The Ecology of Bird Communities*. 2 Volumes. Cambridge: Cambridge University Press.

Wildlife and Countryside (Amendment) Act (1985). Her Majesty's Stationery Office: London.

Williams, G. & Green, R. (1993). Towards an upland habitat action plan. *RSPB Conservation Review* **7**, 5–11.

Williamson, K. (1972*a*). Breeding birds of Ariundle oakwood Forest Nature Reserve. *Quarterly Journal of Forestry* **66**, 243–255.

Williamson, K. (1972*b*). The conservation of bird life in the new coniferous forest. *Forestry* **45**, 87–100.

Williamson, K. (1974). Oak wood breeding bird communities in the Loch Lomond National Nature Reserve. *Quarterly Journal of Forestry* **68**, 9–28.

Williamson, K. (1976). Bird-life in the Wood of Cree, Galloway. *Quarterly Journal of Forestry* **70**, 206–215

Williamson, R. (1978). *The Great Yew Forest – the Natural History of Kingley Vale*. London: Macmillan.

Williamson, R. & Williamson, K. (1973). The bird community of yew woodland at Kingley Vale, Sussex. *British Birds* **66**, 12–23.

Willis, K.J. (1993). How old is ancient woodland? *Trends in Ecology and Evolution* **8**, 427–428.

Wilson, J. (1977). Some breeding bird communities of sessile oak woodlands in Ireland. *Polish Ecological Studies* **3**, 245–256.

Wilson, J. (1978). The breeding bird community of willow scrub at Leighton Moss, Lancashire. *Bird Study* **25**, 239–244.

Wink, M. & Wink, C. (1986). The structure of bird communities of beech-oak woods in relation to exposure, vegetation and altitude. (English title.) *Ökologie der Vögel* **8**, 179–188.

Woodin, S.J. & Farmer, A.M. (1993). Impacts of sulphur and nitrogen deposition on sites and species of nature conservation importance in Great Britain. *Biological Conservation* **63**, 23–30.

Woolhouse, M.E.J. (1983). The theory and practice of the species-area effect, applied to the breeding birds of British woods. *Biological Conservation* **27**, 315–332. (Elsevier Science Ltd., Kidlington, UK.)

Woolhouse, M.E.J. (1987). On species richness and nature reserve design: an empirical study of UK woodland avifauna. *Biological Conservation* **40**, 167–178.

Yapp, W.B. (1959). The birds of the high-level woodlands. The winter population. *Bird Study* **6**, 136–140.

Yapp, W.B. (1962). *Birds and Woods*. Oxford: Oxford University Press.

Yapp, W.B. (1974). Birds of the northwest Highland birchwoods. *Scottish Birds* **8**, 16–31.

Zang, H. (1988). Influence of altitude on the breeding biology of the Nuthatch (*Sitta europaea*) in the Harz Mountains (English title). *Journal für Ornithologie* **129**, 161–174.

Zang, H. (1990). Population decrease of Coal Tit *Parus ater* in the Harz Mountains due to forest damage ('Waldsterben'). (English title.) *Vogelwelt* **111**, 18–28.

INDEX

The Preface and Glossary have not been indexed. Only species names have been indexed from Appendices 1 and 2.

238